Activity Book

CHEMISTRY OF MATTER

Prentice Hall
Englewood Cliffs, New Jersey
Needham, Massachusetts

Activity Book

PRENTICE HALL SCIENCE
Chemistry of Matter

ISBN 0-13-400680-1

8 9 10   97 96 95

Prentice Hall
A Division of Simon & Schuster
Englewood Cliffs, New Jersey 07632

Contents

To the Teacher

The materials in the *Activity Book* are designed to assist you in teaching the *Prentice Hall Science* program. These materials will be especially helpful to you in accommodating a wide range of student ability levels. In particular, the activities have been designed to reinforce and extend a variety of science skills and to encourage critical thinking, problem solving, and discovery learning. The highly visual format of many activities heightens student interest and enthusiasm.

All the materials in the *Activity Book* have been developed to facilitate student comprehension of, and interest in, science. Pages intended for student use may be made into overhead transparencies and masters or used as photocopy originals. The reproducible format allows you to have these items easily available in the quantity you need. All appropriate answers to questions and activities are found at the end of each section in a convenient Answer Key.

CHAPTER MATERIALS

In order to stimulate and increase student interest, the *Activity Book* includes a wide variety of activities and worksheets. All the activities and worksheets are correlated to individual chapters in the student textbook.

Table of Contents

Each set of chapter materials begins with a Table of Contents that lists every component for the chapter and the page number on which it begins. The Table of Contents also lists the number of the page on which the Answer Key for the chapter activities and worksheets begins. In addition, the Table of Contents page for each chapter has a shaded bar running along the edge of the page. This shading will enable you to easily spot where a new set of chapter materials begins.

Whenever an activity might be considered a problem-solving or discovery-learning activity, it is so marked on the Contents page. In addition, each activity that can be used for cooperative-learning groups has an asterisk beside it on the Contents page.

First in the chapter materials is a Chapter Discovery. The Chapter Discovery is best used prior to students reading the chapter. It will enable students to discover for themselves some of the scientific concepts discussed within the chapter. Because of their highly visual design, simplicity, and hands-on approach to discovery learning, the Discovery Activities are particularly appropriate for ESL students in a cooperative-learning setting.

Chapter Activities

Chapter activities are especially visual, often asking students to draw conclusions from diagrams, graphs, tables, and other forms of data. Many chapter activities enable the student to employ problem-solving and critical-thinking skills. Others allow the student to utilize a discovery-learning approach to the topics covered in the chapter. In addition, most chapter activities are appropriate for cooperative-learning groups.

Laboratory Investigation Worksheet

Each chapter of the textbook contains a full-page Laboratory Investigation. A Laboratory Investigation worksheet in each set of chapter materials repeats the textbook Laboratory Investigation and provides formatted space for writing observations and conclusions. Students are aided by a formatted worksheet, and teachers can easily evaluate and grade students' results and progress. Answers to the Laboratory Investigation are provided in the Answer Key following the chapter materials, as well as in the Annotated Teacher's Edition of the textbook.

Answer Key

At the end of each set of chapter materials is an Answer Key for all activities and worksheets in the chapter.

SCIENCE READING SKILLS

Each textbook in *Prentice Hall Science* includes a special feature called the Science Gazette. Each gazette contains three articles.

The first article in every Science Gazette—called Adventures in Science—describes a particular discovery, innovation, or field of research of a scientist or group of scientists. Some of the scientists profiled in Adventures in Science are well known; others are not yet famous but have made significant contributions to the world of science. These articles provide students with firsthand knowledge about how scientists work and think, and give some insight into the scientists' personal lives as well.

Issues in Science is the second article in every gazette. This article provides a nonbiased description of a specific area of science in which various members of the scientific community or the population at large hold diverging opinions. Issues in Science articles introduce students to some of the "controversies" raging in science at the present time. While many of these issues are debated strictly in scientific terms, others involve social issues that pertain to science as well.

The third article in every Science Gazette is called Futures in Science. The setting of each Futures in Science article is some 15 to 150 years in the future and describes some of the advances people may encounter as science progresses through the years. However, these articles cannot be considered "science fiction," as they are all extrapolations of current scientific research.

The Science Gazette articles can be powerful motivators in developing an interest in science. However, they have been written with a second purpose in mind. These articles can be used as science readers. As such, they will both reinforce and enrich your students' ability to read scientific material. In order to better assess the science reading skills of your students, this *Activity Book* contains a variety of science reading activities based on the gazette articles. Each gazette article has an activity that can be distributed to students in order to evaluate their science reading skills.

There are a variety of science reading skills included in this *Activity Book*. These skills include Finding the Main Idea, Previewing, Critical Reading, Making Predictions, Outlining, Using Context Clues, and Making Inferences. These basic study skills are essential in understanding the content of all subject matter, and they can be particularly useful in the comprehension of science materials. Mastering such study skills can help students to study, learn new vocabulary terms, and understand information found in their textbooks.

ACTIVITY BANK

A special feature called the Activity Bank ends each textbook in *Prentice Hall Science*. The Activity Bank is a compilation of hands-on activities designed to reinforce and extend the science concepts developed in the textbook. Each activity that appears in the Activity Bank section of the textbook is reproduced here as a worksheet with space for recording observations and conclusions. Also included are additional activities in the form of worksheets. An Answer Key for all the activities is given. The Activity Bank activities provide opportunities to meet the diverse abilities and interests of students; to encourage problem solving, critical thinking, and discovery learning; to involve students more actively in the learning experience; and to address the need for ESL strategies and cooperative learning.

Contents

*Appropriate for cooperative learning

CHAPTER

Chapter Discovery

Atoms and Bonding

1

Electrons and Energy Levels

Background Information

Electrons are often described as being located in a region outside the atomic nucleus called the electron cloud. The electron cloud is made up of a number of different energy levels. Each energy level can hold only a certain number of electrons. The first energy level, which is closest to the nucleus, can hold only 2 electrons. The second and third levels can each hold up to 8 electrons. The energy levels are always filled in order—for example, an atom with 7 electrons will fill the first energy level with 2, then go on to have 5 electrons in the second level.

When atoms bond with other atoms in chemical reactions, they seek to gain stability by obtaining a full outermost energy level. They can do this by gaining or losing electrons.

Materials

posterboard
colored construction paper
scissors
glue
marking pen

drawing compass
small craft beads (no more
   than several mm in diameter)
10 small paper cups

Procedure

1. On a piece of colored construction paper, draw 10 small circles about the size of a nickel.

2. Label each circle with one of the following:
   He
   Li
   Be
   O
   F
   Ne
   Na
   Mg
   Cl
   Ar

3. Label each of the 10 paper cups with the same abbreviations.

4. Into the cup marked He, drop 2 beads. Into the cup marked Li, drop 3 beads. Continue dropping beads into the cups as follows: Be, 4 beads; O, 8 beads; F, 9 beads; Ne, 10 beads; Na, 11 beads; Mg, 12 beads; Cl, 17 beads; Ar, 18 beads.

5. Cut 10 squares of about 15 cm × 15 cm each from posterboard.

6. Cut out the small circles that you made in steps 1 and 2. These circles represent atomic nuclei. Glue one "nucleus" in the middle of each posterboard square.

7. Around each nucleus, draw a circle about 1 cm from the edge of the nucleus. This will represent the first energy level.

8. Draw two more circles, making each circle about 1 cm away from the previous one. These circles represent the second and third energy levels. When you are finished, you should have a square that looks like Figure 1.

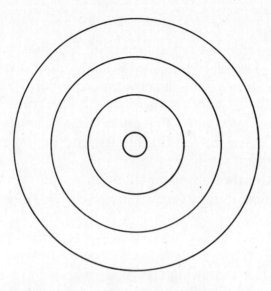

Figure 1

9. Spread out your posterboard squares on a desk or table. Next to each square, place the paper cup that has the same symbol as the nucleus.

10. The beads in each cup represent the electrons found in an atom of that element. Use glue to attach the beads to the correct energy levels. Remember to fill up the innermost (first) energy level first, then move on to the second level. When it becomes full, move on to the third level. As you place the "electrons" in an energy level, try to spread them out evenly around the circle.

11. When all of your squares are finished, open your textbook to Appendix E, which shows the periodic table. Look at the first three rows, or periods, of the table. Arrange your squares on the desk or table in front of you so that they are in the same relative positions as these elements appear in the periodic table. (You may wish to use blank squares to hold the places of elements you did not model in this activity.)

Analysis and Conclusions

1. Look at the elements that appear in the same vertical columns. What do you notice about the outermost energy levels of these elements?

_____

_____

_____

2. Look at the elements that appear in the second and third rows of the table. What happens to the outermost energy levels of the elements as you move across the row from left to right?

_____

_____

_____

3. Which of the elements that you modeled are most likely to lose electrons in a chemical bond? Why?

_____

_____

_____

4. Which of the elements that you modeled are most likely to gain electrons in a chemical bond? Why?

_____

_____

_____

5. Do you think that any of the elements you modeled will not bond with other elements? Which ones? Explain.

_____

_____

_____

_____

6. Predict any chemical bonds that might result between any of the elements that you modeled. Give reasons for your predictions.

_____

_____

_____

_____

Activity

Atomic Structure

Use the information provided for each element to complete the diagrams. Draw the electrons in their proper shells, and place the correct numbers in the nucleus to indicate the number of protons and the number of neutrons.

1. Sulfur: atomic number 16
        atomic mass   32

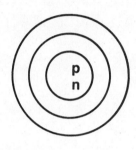

2. Beryllium: atomic number 4
         atomic mass   9

3. Nitrogen: atomic number 7
         atomic mass   14

4. Sodium: atomic number 11
        atomic mass   23

5. Potassium: atomic number 19
         atomic mass   39

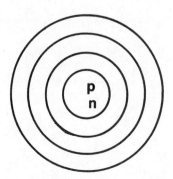

6. Argon: atomic number 18
        atomic mass   40

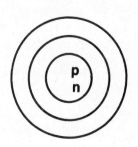

Activity

Classifying Crystals

Part A

Using books and other reference materials in the library, find out about the basic shapes of crystals.

1. Make a poster illustrating these crystal shapes. Give several examples of substances that have each of the crystal shapes.

2. In the space provided, draw a diagram of each crystal shape you have illustrated on your poster. Include the names of the substances you have given as examples of each crystal shape.

Part B

If possible, obtain small samples of crystals or find pictures of crystalline minerals in magazines.

1. Classify each sample according to its crystalline shape and attach the sample to your poster. Display your poster in the classroom.

2. In the space provided, describe the samples or pictures you have found and indicate how you have classified each.

_____

_____

_____

_____

_____

_____

_____

Electron-Dot Diagram

The electron-dot diagram of a hydrogen atom is

H·

The electron-dot diagram of a carbon atom is

·**Ċ**·

Use electron-dot diagrams to show how hydrogen atoms would join with carbon atoms to form at least two covalent compounds. Make sure all the atoms have filled energy levels.

Activity

Drawing Chemical Bonds

Practice making diagrams of ionic and covalent bonds. Draw the bonds formed in the substances listed below. The number of valence electrons in the atoms of each element is provided.

Valence Electrons

F 7	Br 7	Cl 7	Mg 2
H 1	K 1	C 4	

1. Fluorine molecule, F_2

4. Hydrogen bromide, HBr

2. Potassium chloride, KCl

5. Methane, CH_4

3. Magnesium chloride, $MgCl_2$

6. Carbon tetrachloride, CCl_4

A ctivity

Atoms and Bonding

Charting Oxidation Number

Complete the following chart. You may wish to use the periodic table on pages 154 and 155 of the textbook.

Element	Atomic Number	Number of Protons (+)	Number of Electrons (−)	Number of Valence Electrons	Type of Ion Formed	Oxidation Number
Hydrogen	1	1	1	1	+	1+
Helium						
Lithium						
Beryllium						
Boron						
Carbon						
Nitrogen						
Oxygen						
Fluorine						
Neon						
Sodium						
Magnesium						
Aluminum						
Silicon						
Phosphorus						
Sulfur						
Chlorine						
Argon						
Potassium						
Calcium						

Activity

Constructing Molecule Models

1. Obtain 20 Styrofoam balls, about 5 to 10 cm in diameter. Using a quick-drying paint, paint the balls in the following way:

   7 red (hydrogen)
   7 green (chlorine)
   2 yellow (sulfur)
   leave the remaining 4 balls white (oxygen)

2. Use toothpicks to attach the Styrofoam balls to one another. The toothpicks will represent covalent bonds.

3. Construct models of the following molecules: H_2, HCl, Cl_2, O_2, H_2O, attaching both H's to the O at right angles to each other; H_2S, attaching both H's to the S at right angles to each other; OCl_2, attaching both Cl's to the O at right angles to each other; and SCl_2, attaching both Cl's to the S at right angles to each other. Display your models in the classroom.

4. In the space provided, draw a diagram of each molecular model you have constructed. Use colored pencils to indicate the different atoms.

H_2 HCl Cl_2

O_2 H_2O H_2S

OCl_2 SCl_2

Activity _____ **Atoms and Bonding**

CHAPTER

1

Bond Identification

1. When atoms share electrons to fill their outermost energy levels, they form

_____ bonds.

2. When atoms transfer electrons to fill their outermost energy levels, they form

_____ bonds.

3. Indicate whether the atoms listed below will share electrons or transfer electrons.
(*Hint:* When electrons are transferred, both atoms will have complete outermost
energy levels. Otherwise, atoms share electrons.)

    a. $:\overset{..}{\underset{.}{O}}\cdot$ + $\cdot\overset{..}{\underset{.}{O}}:$ _____

    b. $K\cdot$ + $\cdot\overset{..}{\underset{..}{Cl}}:$ _____

    c. $\cdot\overset{.}{\underset{.}{C}}\cdot$ + $\cdot\overset{..}{\underset{.}{O}}:$ _____

    d. $H\cdot$ + $\cdot\overset{..}{\underset{..}{F}}:$ _____

    e. $:\overset{..}{\underset{..}{I}}\cdot$ + $\cdot\overset{..}{\underset{..}{I}}:$ _____

    f. $:\overset{.}{\underset{.}{S}}\cdot$ + $\cdot\overset{..}{\underset{.}{O}}:$ + $\cdot\overset{..}{\underset{.}{O}}:$ _____

    g. $Na\cdot$ + $Na\cdot$ + $:\overset{.}{\underset{.}{S}}\cdot$ _____

    h. $Li\cdot$ + $\cdot\overset{..}{\underset{..}{Cl}}:$ _____

    i. $Ca\cdot$ + $\cdot\overset{..}{\underset{..}{Br}}:$ + $\cdot\overset{..}{\underset{..}{Br}}:$ _____

    j. $\cdot\overset{.}{\underset{..}{N}}:$ + $\cdot\overset{.}{\underset{..}{N}}:$ _____

4. Write the chemical formula and the electron-dot model for each of the products in question 3. If ions are formed, show the electric charge on each.

Chemical Formula	**Electron-Dot Model**

a. _____   _____

b. _____   _____

c. _____   _____

d. _____   _____

e. _____   _____

f. _____   _____

g. _____   _____

h. _____   _____

i. _____   _____

j. _____   _____

Activity _____ CHAPTER

Chemical Analogies

Choose one of the three words below the line that relates to the single word in the same way as the first two words relate to each other. Make sure you can explain your answer choice.

Example

physical : chemical appearance: ___reactivity_____
reactivity, taste, solubility

Reason: physical property vs. chemical property

1. element : compound oxygen: _____
water, hydrogen, matter

2. compound : mixture chemical: _____
physical, separation, property

3. hydrogen : water carbon: _____
graphite, coal, carbon dioxide

4. combine : decompose ionic: _____
mixture, covalent, atom

5. gas : condensing solid: _____
melting, freezing, decomposing

6. ionic : covalent ions: _____
binary, molecule, polyatomic

7. iron : rust silver: _____
reaction, oxygen, tarnish

8. oxygen : gas iron: _____
rust, tarnish, solid

9. chemical : rust physical: _____
compound, condensation, solid

10. ionic : transfer covalent: _____
molecule, share, ion

11. halogens : reactive noble gas: _____
solid, stable, inert

12. oxygen : molecule

chloride: _____
ion, halogen, binary

13. ratio : subscript

oxidation number: _____
charge, bond, transfer

Activity

Bonding and Chemical Formulas

When a chemical formula for a compound is written correctly, it shows the number of each type of atom in the compound. These numbers, called subscripts, are determined by the bonding between the atoms.

The table shows two columns of elements. The elements in the first column usually give up electrons when they form compounds. The elements in the second column usually gain electrons when they form compounds. The column next to the elements gives the number of electrons found in the outer level of each element. Using this information, determine the charge on the ion after the exchange of electrons. Remember, atoms that give up electrons become positive ions, while atoms that gain electrons become negative ions.

Now show how the positive ion would combine with the negative ion to make a neutral compound. For example, sodium, Na, has 1 electron in its outer level. It gives up this electron and becomes a 1+ ion. Sulfur, S, with 6 electrons in its outer level, gains 2 electrons to fill this outer level with 8 electrons. Sulfur becomes a 2− ion. These 2 ions then combine to form Na_2S. This formula is correct because it takes 2 sodium ions to match the 2− charge on 1 sulfur ion.

Element	Electrons in Outer Level	Charge on the Ion	Element	Electrons in Outer Level	Charge on the Ion	Formula
Aluminum	3		Chlorine	7		
Magnesium	2		Bromine	7		
Sodium	1		Oxygen	6		
Lithium	1		Oxygen	6		
Calcium	2		Phosphorus	5		
Carbon	4		Chlorine	7		
Aluminum	3		Oxygen	6		
Beryllium	2		Sulfur	6		
Sodium	1		Fluorine	7		
Silicon	4		Neon	8		

_____ *Laboratory Investigation* _____

Properties of Ionic and Covalent Compounds

Problem
Do covalent compounds have different properties from ionic compounds?

Materials *(per group)*

safety goggles	distilled water
salt	(200 mL)
4 medium-sized	2 100-mL
test tubes	beakers
glass-marking	stirring rod
pencil	3 connecting
test-tube tongs	wires
Bunsen burner	light bulb socket
timer	light bulb
sugar	dry-cell battery
vegetable oil	

Procedure 🜂 ♨ 👆 👁 ·❙⊦

1. Place a small sample of salt in a test tube. Label the test tube. Place an equal amount of sugar in another test tube. Label that test tube.

2. Using tongs, heat the test tube of salt over the flame of the Bunsen burner. **CAUTION:** *Observe all safety precautions when using a Bunsen burner.* Determine how long it takes for the salt to melt. Immediately stop heating when melting begins. Record the time.

3. Repeat step 2 using the sugar.

4. Half fill a test tube with vegetable oil. Place a small sample of salt in the test tube. Shake the test tube gently for about 10 seconds. Observe the results.

5. Repeat step 4 using the sugar.

6. Pour 50 mL of distilled water into a 100-mL beaker. Add some salt and stir until it is dissolved. To another 100-mL beaker add some sugar and stir until dissolved.

7. Using the beaker of salt water, set up a circuit as shown on page 32. **CAUTION:** *Exercise care when using electricity.* Observe the results. Repeat the procedure using the beaker of sugar water.

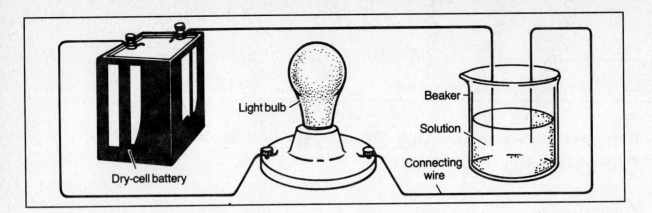

Light bulb

Beaker

Solution

Connecting
wire

Dry-cell battery

Observations

1. Does the salt or the sugar take a longer time to melt?

_____

2. Does the salt dissolve in the vegetable oil? Does the sugar?

_____

3. Which compound is a better conductor of electricity?

_____

Analysis and Conclusions

1. Which substance do you think has a higher melting point? Explain.

_____

_____

_____

_____

2. Explain why one compound is a better conductor of electricity than the other.

_____

_____

_____

_____

3. How do the properties of each type of compound relate to their bonding?

_____

_____

_____

_____

Answer Key

Chapter Discovery: Electrons and Energy Levels **1.** Elements in the same column have the same number of electrons in the outermost level. **2.** The number of electrons in the outermost energy level increases as you move from left to right across a period. **3.** Li, Be, Na, and Mg because they have only 1 or 2 electrons in the outermost level. **4.** O, F, Cl because they need only 1 or 2 electrons to complete the outermost level. **5.** He, Ne, and Ar because their outermost energy levels are already full. **6.** Accept all reasonable predictions. Students should figure out that elements that have 1 electron in the outer level may bond with elements that have 7 electrons in the outer level and so on.

Activity: Atomic Structure
1. 16p, 16n **2.** 4p, 5n **3.** 7p, 7n **4.** 11p, 12n **5.** 19p, 20n **6.** 18p, 22n

Discovery Activity: Classifying Crystals
Part A There are seven crystal shapes: cubic, tetragonal, hexagonal, trigonal, orthorhombic, monoclinic, triclinic. Students should be able to find illustrations of these shapes in a high school chemistry textbook.
Part B Answers will vary, depending on the samples students were able to locate.

Problem-Solving Activity: Electron-Dot Diagram
Two examples are:

$$CH_4 \text{ methane } \quad \ddot{H} \\ H \colon \ddot{C} \colon H \\ \ddot{H}$$

4 atoms of hydrogen, 1 atom of carbon

$$C_2H_6 \text{ ethane } \quad H\ H \\ H \colon \ddot{C} \colon \ddot{C} \colon H \\ H\ H$$

6 atoms of hydrogen, 2 atoms of carbon

Activity: Drawing Chemical Bonds

1. $F \colon F$ **4.** $H \colon \ddot{Br} \colon$

2. $K \colon \ddot{Cl} \colon$ **5.** $H \colon \ddot{Cl} \colon H$ with H above and below

3. $Mg \begin{array}{c} \colon \ddot{Cl} \colon \\ \colon \ddot{Cl} \colon \end{array}$ **6.** $\colon \ddot{Cl} \colon \ddot{C} \colon \ddot{Cl} \colon$ with $\colon \ddot{Cl} \colon$ above and below

Activity: Charting Oxidation Number
hydrogen 1, 1, 1, 1, +, 1+ **helium** 2, 2, 2, 2, 0, 0, **lithium** 3, 3, 3, 1, +, 1+ **beryllium** 4, 4, 4, 2, +, 2+ **boron** 5, 5, 5, 3, +, 3+ **carbon** 6, 6, 6, 4, + or −, 4 +, 4− **nitrogen** 7, 7, 7, 5, −, 3− **oxygen** 8, 8, 8, 6, −, 2− **fluorine** 9, 9, 9, 7, −, 1− **neon** 10, 10, 10, 8, 0, 0 **sodium** 11, 11, 11, 1, +, 1+ **magnesium** 12, 12, 12, 2, +, 2+ **aluminum** 13, 13, 13, 3, +, 3+ **silicon** 14, 14, 14, 4, + or −, 4+, 4− **phosphorus** 15, 15, 15, 5, −, 3− **sulfur** 16, 16, 16, 6, −, 2− **chlorine** 17, 17, 17, 7, −, 1− **argon** 18, 18, 18, 8, 0, 0 **potassium** 19, 19, 19, 1, +, 1+ **calcium** 20, 20, 20, 2, +, 2+

Activity: Constructing Molecule Models
Models may be displayed by suspending them from a string or they may be mounted on a ball of modeling clay.

Activity: Bond Identification
1. covalent **2.** ionic **3a.** share **b.** transfer **c.** share **d.** share **e.** share **f.** share **g.** transfer **h.** transfer **i.** transfer **j.** share

4a. O_2 $O \colon\colon O$

b. KCl $K^+ \ \colon \ddot{Cl} \colon^-$

c. CO   :C:::O:

d. HF   H:F̈:

e. I_2 :Ï::Ï:

f. SO_2
$$\overset{\displaystyle S}{\underset{:\ddot{O}:}{\ddot{}\ }}\ \ddot{O}:$$

g. Na_2S $2Na^+:\ddot{S}:^{-2}$

h. LiCl $Li^+:\ddot{Cl}:^-$

i. $CaBr_2$ $Ca^{+2}\ 2:\ddot{Br}:^{-2}$

j. N_2 :N:::N:

Problem Solving Activity: Chemical Analogies

1. water   **2.** physical   **3.** carbon dioxide
4. covalent   **5.** melting   **6.** molecule
7. tarnish   **8.** solid   **9.** condensation
10. share   **11.** stable   **12.** ion   **13.** charge

Activity: Bonding and Chemical Formulas

Aluminum 3 +3	Chlorine 7 − 1	$AlCl_3$
Magnesium 2 +2	Bromine 7 −1	$MgBr_2$

Sodium 1 +1	Oxygen 6 −2	Na_2O
Lithium 1 +1	Oxygen 6 −2	Li_2O
Calcium 2 +2	Phosphorus 5 −3	Ca_3P_2
Carbon 4 +4	Chlorine 7 −1	CCl_4
Aluminum 3 +3	Oxygen 6 −2	Al_2O_3
Beryllium 2 +2	Sulfur 6 −2	BeS
Sodium 1 +1	Fluorine 7 −1	NaF
Silicon 4 +4	Neon 8 0	no compound

Laboratory Investigation: Properties of Ionic and Covalent Compounds

Observations   1. Salt.   **2.** No; yes.   **3.** Salt is the better conductor because it forms ions when dissolved in water.   **Analysis and Conclusions   1.** Salt. It took longer for the salt to melt, if it did in fact melt.   **2.** The crystal lattice structure of an ionic compound lends itself to conducting electricity better than the more separate units of a covalent bond do.   **3.** Covalent compounds typically have low melting points, dissolve in covalent solvents, and do not conduct an electric current. Ionic compounds typically have high melting points, do not dissolve in covalent solvents, and conduct electricity when dissolved in water.

Contents

CHAPTER 2

*Appropriate for cooperative learning

Chapter Discovery

Chemical Reactions

Kitchen Chemistry

Background Information

A chemical reaction is a process in which the physical and chemical properties of a substance or substances change as a new substance or substances with different physical and chemical properties are formed.

Materials

3 dishes
masking tape
marking pen
table salt (sodium chloride)
baking soda (sodium bicarbonate)
powdered laundry starch
vinegar (acetic acid)
water
iodine solution
3 medicine droppers
piece of table silver
1 egg
paper towels

Procedure 🔥 🧰

1. Attach a small piece of masking tape to each of three dishes or plates. On the tape, label the plates A, B, and C.

2. On plate A, arrange three piles of table salt. Each pile should be a few centimeters in diameter. Leave as much space as possible between the piles.

3. Repeat step 2 on plate B using baking soda and on plate C using laundry starch.

4. Fill a medicine dropper with vinegar.

5. Place several drops of vinegar on one pile in each of the three plates. Record any changes that you observe.

6. Fill another medicine dropper with iodine solution. Place several drops of iodine on one of the two remaining piles in each of the three plates. Record your observations.

7. Fill a third medicine dropper with water. Place several drops of water on the remaining pile in each of the three plates. Record your observations.

8. Break open an egg and separate the yolk from the white. Dip one end of a silver table utensil in the egg yolk. **CAUTION:** *Be sure that you have the permission of an adult in your home before using a piece of silver for this experiment.* Rub the egg yolk evenly over the end of the utensil. Then place the utensil on several sheets of paper toweling for 10 to 15 minutes.

9. Gently wash the piece of silver. What changes do you see in the silver? Record your observations.

Observations

Sample	Substance	Observations
Salt	Vinegar	
Salt	Iodine	
Salt	Water	
Baking soda	Vinegar	
Baking soda	Iodine	
Baking soda	Water	
Starch	Vinegar	
Starch	Iodine	
Starch	Water	
Silver	Egg yolk	

Critical Thinking and Application

1. Which substances caused a change in table salt? Were these physical or chemical changes?

_____

_____

2. Consider your data for baking soda. Do any of your observations make you suspect that a chemical change has taken place? If so, why?

_____

_____

3. What changes did you observe in the starch? Were they physical or chemical changes? Explain the reasoning for your answer.

_____

_____

_____

4. What happened when you placed the silver utensil in the egg yolk? Do you think a chemical reaction occurred? Explain.

_____

_____

Activity

Chemical Reactions

Testing Reaction Rate Factors

Devise and conduct an experiment to see what effect temperature, stirring, and the size of particles have on the rate at which sugar dissolves in water. Write up your experiment using standard form: Problem, Materials, Procedure, Observations, Conclusions. Use diagrams and data tables when appropriate. Identify your control and variable for each item being tested.

Problem

Materials

Procedure

Observations

Conclusions

— **A**ctivity ——————————————————————————— CHAPTER
                                    Chemical Reactions **2**

ctivity

Conservation of Mass

Observe some common changes in matter such as ice melting; water freezing or boiling; leaves, wood, or matches burning; sugar dissolving in beverages; and so on. In each case, describe the matter before and after the change. Then answer these questions.

1. What properties of matter seem to be changing?

_____

_____

2. Does it seem that the law of conservation of mass is operating in each change? Why?

_____

_____

_____

_____

_____

_____

_____

_____

3. What makes it difficult to observe this law in operation? Describe an experiment you could perform to show whether mass is conserved in each case.

_____

_____

_____

_____

_____

_____

_____

_____

Activity

Use only Coefficients.   **Balancing Equations**   *Do Not Change Formulas or Subscripts*

Start over again at beginning after each Coefficient Change.

1. $Na + O_2 \rightarrow Na_2O$

2. $H_2 + O_2 \rightarrow H_2O$

3. $Na_2SO_4 + CaCl_2 \rightarrow CaSO_4 + NaCl$

4. $Al_2O_3 \rightarrow Al + O_2$

5. $N_2 + H_2 \rightarrow NH_3$

6. $Fe + H_2O \rightarrow Fe_3O_4 + H_2$

7. $P_4 + O_2 \rightarrow P_4O_{10}$

8. $C_2H_6 + O_2 \rightarrow CO_2 + H_2O$

9. $SiCl_4 \rightarrow Si + Cl_2$

10. $C + H_2 \rightarrow CH_4$

ctivity

Types of Chemical Reactions

1. A decomposition reaction starts with one reactant and ends up with two or more products. Which of the following reactions are decomposition reactions? Circle the letters.
   a. $NaCl \rightarrow Na + Cl_2$
   b. $Na + Cl_2 \rightarrow NaCl$
   c. $H_2O \rightarrow H_2 + O_2$
   d. $H_2 + O_2 \rightarrow H_2O$
   e. $NaOH + HCl \rightarrow HOH + NaCl$

2. A synthesis reaction starts with two reactants and ends up with one product. Which of the following reactions are synthesis reactions? Circle the letters.
   a. $NaCl \rightarrow Na + Cl_2$
   b. $Na + HCl \rightarrow H_2 + NaCl$
   c. $H_2 + O_2 \rightarrow H_2O$
   d. $NaOH + HCl \rightarrow HOH + NaCl$
   e. $K + Cl_2 \rightarrow KCl$

3. A single-replacement reaction starts with two reactants and ends up with two products. The uncombined element takes the place of the combined element in the compound. Which of the following reactions are single-replacement reactions? Circle the letters.
   a. $NaCl \rightarrow Na + Cl_2$
   b. $NaOH + HCl \rightarrow HOH + NaCl$
   c. $K + AgCl \rightarrow Ag + KCl$
   d. $Ca + S \rightarrow CaS$
   e. $Na + HCl \rightarrow H_2 + NaCl$

4. A double-replacement reaction starts with two reactants and ends up with two products. In this case both reactants are compounds and both products are compounds. They simply change partners. Which of the following reactions are double-replacement reactions? Circle the letters.
   a. $NaCl \rightarrow Na + Cl_2$
   b. $NaOH + HCl \rightarrow HOH + NaCl$
   c. $Na + HCl \rightarrow H_2 + NaCl$
   d. $KOH + HNO_3 \rightarrow HOH + KNO_3$
   e. $Ca + S \rightarrow CaS$

ctivity

Chemical Reactions

CHAPTER

2

Completing Equations

Chemical changes are characterized by the formation of new substances. The products of a correctly balanced chemical equation represent the number and kind of new substances formed. There are four general types of reactions: synthesis, decomposition, single replacement, and double replacement.

Examine each equation listed below and identify which type of reaction is taking place by filling in the blank space to the right of the equation.

Then balance the equation. Remember to use coefficients, not subscripts, to balance the equation.

1. $Al + O_2 \rightarrow Al_2O_3$ _____

2. $HgO \rightarrow Hg + O_2$ _____

3. $NaOH + H_2SO_4 \rightarrow Na_2SO_4 + H_2O$ _____

4. $Fe + O_2 \rightarrow Fe_2O_3$ _____

5. $Pb(NO_3)_2 + K_2CrO_4 \rightarrow PbCrO_4 + KNO_3$ _____

6. $H_2 + N_2 \rightarrow NH_3$ _____

7. $C_3H_5(NO_3)_3 \rightarrow CO_2 + N_2 + H_2O + O_2$ _____

8. $Fe + CuCl_2 \rightarrow FeCl_2 + Cu$ _____

9. $KClO_3 \rightarrow KCl + O_2$ _____

10. $Mg + HCl \rightarrow H_2 + MgCl_2$ _____

Activity

Identifying and Balancing Chemical Equations

Identify each of the equations below as synthesis, decomposition, single replacement, or double replacement.

1. $HgO \rightarrow Hg + O_2$ _____

2. $NaCl + AgNO_3 \rightarrow NaNO_3 + AgCl$ _____

3. $Mg + HCl \rightarrow MgCl_2 + H_2$ _____

4. $Zn + H_2SO_4 \rightarrow ZnSO_4 + H_2$ _____

5. $NaOH + HCl \rightarrow NaCl + H_2O$ _____

6. $Al_2(SO_4)_3 + Ca(OH)_2 \rightarrow Al(OH)_3 + CaSO_4$ _____

7. $H_2 + O_2 \rightarrow H_2O$ _____

8. $Cl_2 + NaBr \rightarrow NaCl + Br_2$ _____

9. $Zn + CuSO_4 \rightarrow ZnSO_4 + Cu$ _____

10. $KClO_3 \rightarrow KCl + O_2$ _____

11. $H_2O + Fe \rightarrow Fe_2O_3 + H_2$ _____

12. $Ca(OH)_2 + HNO_3 \rightarrow Ca(NO_3)_2 + H_2O$ _____

13. $Na_2O + CO_2 \rightarrow Na_2CO_3$ _____

14. $H_2 + N_2 \rightarrow NH_3$ _____

Balance the following chemical equations.

15. $HgO + Cl_2 \rightarrow HgCl + O_2$

16. $Na + Br_2 \rightarrow NaBr$

17. $KClO_3 \rightarrow KCl + O_2$

18. $Ca(OH)_2 + HNO_3 \rightarrow Ca(NO_3)_2 + H_2O$

19. $Al_2O_3 \rightarrow Al + O_2$

20. $CuCl_2 + H_2S \rightarrow CuS + HCl$

21. $Cl_2 + NaBr \rightarrow NaCl + Br_2$

22. $NaOH + HCl \rightarrow NaCl + H_2O$

23. $Na_2O + CO_2 \rightarrow Na_2CO_3$

24. $H_2O + Fe \rightarrow Fe_2O_3 + H_2$

Activity

Chemical Formulas and Equations

Write the chemical formula for each of the following.

1. potassium chloride

2. sodium hydroxide

3. calcium oxide

4. calcium fluoride

5. silicon dioxide

6. aluminum oxide

7. silver nitrate

8. calcium carbonate

9. magnesium nitrate

10. sodium bicarbonate

Balance each of the following equations.

11. $P_4 + O_2 \rightarrow P_4O_{10}$

12. $Zn + HCl \rightarrow ZnCl_2 + H_2$

13. $Mg + O_2 \rightarrow MgO$

14. $KClO_3 \rightarrow KCl + O_2$

15. $FeS_2 + O_2 \rightarrow Fe_2O_3 + SO_2$

Write a balanced equation for each of the following.

16. Silver + Chlorine $\rightarrow$ Silver chloride

17. Hydrogen + Chlorine $\rightarrow$ Hydrogen chloride

18. Hydrochloric acid + Sodium hydroxide $\rightarrow$ Sodium chloride + Water

19. Magnesium hydroxide + Hydrochloric acid $\rightarrow$ Magnesium chloride + Water

20. Calcium carbonate $\rightarrow$ Calcium oxide + Carbon dioxide

Name the types of reactions involved in questions 11 through 20 (synthesis, decomposition, single replacement, or double replacement).

21. (11) _____

22. (12) _____

23. (13) _____

24. (14) _____

25. (15) _____

26. (16) _____

27. (17) _____

28. (18) _____

29. (19) _____

30. (20) _____

Laboratory Investigation

Determining Reaction Rate

Problem
How does concentration affect reaction rate?

Materials (*per group*)
safety goggles
2 graduated cylinders
120 mL Solution A
3 250-mL beakers
distilled water at room temperature
90 mL solution B
stirring rod
sheet of white paper
stopwatch or watch with a sweep second hand

Procedure 🜄 🔬 👁

1. Carefully measure 60 mL of Solution A and pour it into a 250-mL beaker. Add 10 mL of distilled water and stir.

2. Carefully measure 30 mL of Solution B and pour it into second beaker. Place the beaker of Solution B on a sheet of white paper in order to see the color change more easily.

3. Add the 70 mL of Solution A-water mixture to Solution B. Stir rapidly. Record the time it takes for the reaction to occur.

4. Rinse and dry the reaction beaker.

5. Repeat the procedure using the other amounts shown in the data table.

Observations

Solution A (mL)	Distilled Water Added to Solution A (mL)	Solution B (mL)	Reaction Time (sec)
60	10	30	
40	30	30	
20	50	30	

1. What visible indication is there that a chemical reaction is occurring?

_____

_____

2. What is the effect of adding more distilled water on the concentration of Solution A?

_____

_____

3. What happens to reaction time as more distilled water is added to Solution A?

_____

_____

4. Make a graph of your observations by plotting time along the X axis and volume of Solution A along the Y axis.

Analysis and Conclusions

1. How does concentration affect reaction rate?

_____

_____

2. Does your graph support your answer to question 1? Explain why.

_____

_____

3. What would a graph look like if time were plotted along the X axis and volume of distilled water added to Solution A were plotted along the Y axis?

_____

_____

4. In this investigation, what is the variable? The control?

_____

_____

5. On Your Own   Enhance your graph by testing other concentrations.

Answer Key

Chapter Discovery: Kitchen Chemistry

Observations Salt: no reaction; reddish color; dissolves. Baking soda: bubbles; absorbs color; bubbles. Starch: milky mixture; turns black; milky mixture. Silver: silver turns black. **Critical Thinking and Application 1.** Iodine and water; both changes are physical. **2.** The reactions with vingear and water produce gas bubbles. The bubbles must be a new substance that was not there originally. In both cases, a chemical reaction has occurred, producing carbon dioxide. **3.** The starch changed when all three substances were added. All changes are physical changes. The reactions with water and vinegar form solutions. The reaction with iodine causes a color change. Neither the starch nor the iodine changes into a new substance. **4.** Students may think that because the color—a physical property—has changed, only a physical change has taken place. However, the egg yolk is completely gone and the silver has lost its shine. A chemical reaction has occurred in which sulfur in the egg yolk has combined with silver to form silver sulfide.

Problem-Solving Activity: Testing Reaction Rate Factors

Answers will vary. Make sure that the approach the students take is logical. Make sure that the experiment they design follows the form of investigations in the textbook and has a control.

Discovery Activity: Conservation of Mass

1. During phase changes, the mass remains the same but the density and volume change. **2.** If the mass of each object were measured before and then after the chemical change had taken place, the mass would be found to be the same. **3.** It would be difficult to calculate the mass of water when it was converted from ice to steam, although this could be done by using a distillation process, provided that no steam escaped. To show the conservation of mass as ice melts to water, weigh a graduated cylinder. Place a few ice cubes in the cylinder and weigh it. By finding the difference in mass between the empty cylinder and the cylinder containing ice, the mass of the ice can be determined. After the ice has melted, weigh the cylinder that contains the water. Compare the two masses to illustrate how mass is conserved as a substance changes phases.

Problem-Solving Activity: Balancing Equations

1. $4Na + O_2 \rightarrow 2Na_2O$ **2.** $2H_2 + O_2 \rightarrow 2H_2O$ **3.** $Na_2SO_4 + CaCl_2 \rightarrow CaSO_4 + 2NaCl$ **4.** $2Al_2O_3 \rightarrow 4Al + 3O_2$ **5.** $N_2 + 3H_2 \rightarrow 2NH_3$ **6.** $3Fe + 4H_2O \rightarrow Fe_3O_4 + 4H_2$ **7.** $P_4 + 5O_2 \rightarrow P_4O_{10}$ **8.** $2C_2H_6 + 7O_2 \rightarrow 4CO_2 + 6H_2O$ **9.** $SiCl_4 \rightarrow Si + 2Cl_2$ **10.** $C + 2H_2 \rightarrow CH_4$

Activity: Types of Chemical Reactions

1. a, c **2.** c, e **3.** c, e **4.** b, d

Activity: Completing Equations

1. Synthesis: $4Al + 3O_2 \rightarrow 2Al_2O_3$
2. Decomposition: $2HgO \rightarrow 2Hg + O_2$
3. Double Replacement: $2NaOH + H_2SO_4 \rightarrow Na_2SO_4 + 2H_2O$
4. Synthesis: $4Fe + 3O_2 \rightarrow 2Fe_2O_3$
5. Double Replacement: $Pb(NO_3)_2 + K_2CrO_4 \rightarrow PbCrO_4 + 2KNO_3$
6. Synthesis: $3H_2 + N_2 \rightarrow 2NH_3$
7. Decomposition: $4C_3H_5(NO_3)_3 \rightarrow 12CO_2 + 6N_2 + 10H_2O + O_2$
8. Single Replacement: $Fe + CuCl_2 \rightarrow FeCl_2 + Cu$
9. Decomposition: $2KClO_3 \rightarrow 2KCl + 3O_2$
10. Single Replacement: $Mg + 2HCl \rightarrow MgCl_2 + H_2$

Problem-Solving Activity: Identifying and Balancing Chemical Equations

1. decomposition   2. double replacement
3. single replacement   4. single replacement
5. double replacement   6. double replacement   7. synthesis   8. single replacement   9. single replacement
10. decomposition   11. single replacement
12. double replacement   13. synthesis
14. synthesis
15. $2HgO + Cl_2 \rightarrow 2HgCl + O_2$
16. $2Na + Br_2 \rightarrow 2NaBr$
17. $2KClO_3 \rightarrow 2KCl + 3O_2$
18. $Ca(OH)_2 + 2HNO_3 \rightarrow Ca(NO_3)_2 + 2H_2O$
19. $2Al_2O_3 \rightarrow 4Al + 3O_2$
20. $CuCl_2 + H_2S \rightarrow CuS + 2HCl$
21. $Cl_2 + 2NaBr \rightarrow 2NaCl + Br_2$
22. $NaOH + HCl \rightarrow NaCl + H_2O$
23. $Na_2O + CO_2 \rightarrow Na_2CO_3$
24. $3H_2O + 2Fe \rightarrow Fe_2O_3 + 3H_2$

Activity: Chemical Formulas and Equations

1. KCl 2. $NaOH$ 3. CaO 4. CaF_2
5. SiO_2 6. Al_2O_3 7. $AgNO_3$ 8. $CaCO_3$
9. $Mg(NO_3)_2$
10. $NaHCO_3$
11. $P_4 + 5O_2 \rightarrow P_4O_{10}$
12. $Zn + 2HCl \rightarrow ZnCl_2 + H_2$
13. $2Mg + O_2 \rightarrow 2MgO$
14. $2KClO_3 \rightarrow 2KCl + 3O_2$
15. $4FeS_2 + 11O_2 \rightarrow 2Fe_2O_3 + 8SO_2$
16. $2Ag + Cl_2 \rightarrow 2AgCl$

17. $H_2 + Cl_2 \rightarrow 2HCl$
18. $HCl + NaOH \rightarrow NaCl + H_2O$
19. $Mg(OH)_2 + 2HCl \rightarrow MgCl_2 + 2H_2O$
20. $CaCO_3 \rightarrow CaO + CO_2$
21. synthesis   22. single replacement
23. synthesis   24. decomposition   25. single replacement   26. synthesis   27. synthesis
28. double replacement   29. double replacement   30. decomposition

Laboratory Investigation: Determining Reaction Rate

Observations 1. The color change is a visible indication that a chemical reaction is taking place.   **2.** By adding more distilled water, the concentration is lessened.   **3.** As more distilled water is added, the reaction time increases because the reaction rate is decreasing.   **4.** Check student's graphs. They should show a straight line going from the bottom-left corner toward the top-right corner.   **Analysis and Conclusions**
1. Decreased concentration means decreased reaction rate, which would show up as increased reaction time.   **2.** Students' graphs should support their answer.   **3** The graph would look the same.   **4.** The variable is the concentration of Solution A. The control is the concentration of Solution B.
5. Students' graphs will vary, depending on concentrations tested. But in general, the graphs should have the same basic shape.

Contents

*Appropriate for cooperative learning

Chapter Discovery   Families of Chemical Compounds

What's in a Flavor?

Background Information

Many artificial flavorings are esters. Esters are compounds that are made by reacting an organic acid with an alcohol. Esters are named according to the alcohol and acid that produce them. For example, ethyl alcohol and benzoic acid combine to form ethyl benzoate. Methyl alcohol and butyric acid combine to form methyl butyrate.

Materials

isoamyl alcohol	test-tube rack	marking pen
methyl alcohol (methanol)	test-tube holder (tongs)	apricot nectar
ethyl alcohol (ethanol)	heat source	pineapple juice or
butyric acid	500-mL beaker	canned pineapple
salicylic acid	water	grape juice
decanoic acid	aluminum foil	gum or candy flavored
sulfuric acid	graduated cylinder	with wintergreen
4 test tubes	masking tape	4 small paper cups

Procedure 🔬 🔥 📦 ✋ 👁

CAUTION: *This activity requires you to work with strong acids and other chemicals. Before you begin, put on your safety goggles and laboratory apron.*

1. Pour apricot nectar into one paper cup, pineapple juice into a second cup, and grape juice into a third cup. Place wintergreen gum or candy in a fourth cup. Note the odor of each substance.

2. Place four test tubes in a test-tube rack. Using small pieces of masking tape, identify the tubes as A, B, C, and D.

3. Fill a 500-mL beaker three-fourths full with water. Place the beaker in contact with a heat source. Heat the water to boiling. Adjust the heat so that the water continues to boil at a moderate rate. **CAUTION:** *Observe all safety rules for working with heat.*

4. Place 2 mL of isoamyl alcohol in test tube A.

5. Place 2 mL of ethyl alcohol in test tubes B and D.

6. Place 2 mL of methyl alcohol in test tube C.

7. Add 2 mL of butyric acid to test tubes A and B.

8. Add 2 mL of salicylic acid to test tube C.

9. Add 2 mL of decanoic acid to test tube D.

10. Add 1 mL of sulfuric acid to each test tube. **CAUTION:** *Be very careful when handling acid. Follow all safety rules when working with acid. Do not allow any acid to splash on your skin or clothing.*

11. Gently tap the bottom of each test tube with your finger to mix the chemicals. **CAUTION: Do not shake the tubes.**

12. Place test tube A in the beaker of boiling water. Allow it to heat for several minutes. Use tongs to remove the test tube from the water. Use your free hand to waft some vapors from the open tube toward your nose. **CAUTION:** *Do not try to sniff the test tube directly.* Note the odor. Does it resemble any of the odors that you sampled in step 1? Record you observations.
(If you do not detect any odor, cover the test tube with aluminum foil and set it aside in a warm place. Check the test tube periodically to see if an odor has developed.)

13. Repeat step 12 with each of the three remaining test tubes.

Observations

Test Tube	Alcohol	Acid	Odor
A	Isoamyl	Butyric	
B	Ethyl	Butyric	
C	Methyl	Salicylic	
D	Ethyl	Decanoic	

Critical Thinking and Application

1. If you wanted to manufacture pineapple ice cream with artificial flavoring, what chemicals would you use?

   _____

2. What chemicals would you need to create chocolate cream candies with apricot-flavored filling?

   _____

3. What chemicals give wintergreen-flavored toothpaste its artificial flavoring?

   _____

   _____

4. What chemicals would you expect to find in a factory that produces grape-flavored bubble gum?

   _____

5. Based on the information at the beginning of this activity, see if you can name the esters that provide each of the flavors in this activity.

   _____

   _____

   _____

Activity

Names and Uses of Salts

Use reference materials in your library to learn more about some widely used salts. Find out the common name, if there is one, and the use for each of the following.

Name	Common Name	Use
Sodium bicarbonate		
Potassium nitrate		
Sodium nitrate		
Sodium carbonate		
Silver nitrate		
Copper sulfate		
Potassium bromide		

Activity _____ **Families of Chemical Compounds**   CHAPTER **3**

Acid, Base, or Neutral Substance?

Each of the following drawings represents a substance that can be classified as an acid, a base, or a neutral substance. On the blank lines provided, indicate which of these the substance is. Also indicate what its pH would likely be using the descriptions 7, above 7, below 7.

1. a. _____   **2.** a. _____   **3.** a. _____

   b. _____   b. _____   b. _____

4. a. _____   **5.** a. _____   **6.** a. _____

   b. _____   b. _____   b. _____

Activity

Dissolving Household Substances

Experiment to find out how quickly and easily various solid substances found in your home dissolve in water. You may try sugar, flour, powdered drinks, cornstarch, instant coffee, talcum powder, soap powder, and so on.

1. Crush whatever materials you have chosen into pieces of about the same size.

2. Find out how much of each substance you can dissolve in samples of a certain amount of water. If the mixture remains cloudy instead of becoming clear, you have not made a true solution.

3. Record your findings in the Data Table.

Amount of water used for each sample: _____

DATA TABLE

Substance Tested					
Amount Used					

Activity

Families of Chemical Compounds

Acid, Base, or Salt?

The following Data Table contains a list of compounds that can be classified as acids, bases, or salts.

1. Decide what type of substance each compound is and fill in the appropriate space in the column. Base your answer on the compound's formula and on the one piece of information supplied about either its pH or its reaction with an indicator.

2. Fill in any remaining blank spaces for that compound.

To help you become familiar with these substances, information about the use of each compound has been provided. You will probably find this information quite interesting.

Compound	Acid, Base, or Salt	Color in Litmus	Color in Phenolphthalein	Approximate pH
H_3PO_4 found in some cola beverages		red		
$Ca(OH)_2$ slaked lime used for treating soil				above 7
$CaSO_4$ used in plaster of Paris				about 7
H_2CO_3 produced in human body from metabolism of fats and sugars			clear	
Na_3PO_4 used in some water softeners		no effect		
H_2S odor of rotten eggs				below 7
H_2SO_3 component in acid rain		red		
$Mg(OH)_2$ found in milk of magnesia		blue		
NH_4Cl used in batteries and dry cells				about 7 (slightly below)
$(NH_4)_2SO_4$ nitrogen fertilizer			clear	

Activity   **Families of Chemical Compounds**   CHAPTER **3**

Digestion

1. Put one or two soda crackers in your mouth. Use unsalted ones if you have them. Crackers contain a great deal of starch. Notice their taste.

2. Now begin to chew the crackers. Try to keep chewing without swallowing for about a minute.

What changes, if any, do you notice in the taste of the food? How would you explain what you have observed?

_____

_____

_____

_____

_____

_____

_____

_____

_____

_____

_____

_____

_____

_____

_____

_____

_____

_____

Use reference materials in the library to find out more about the enzymes that are important for digestion of the food you eat. Find out where in your body each of the following enzymes is produced and what kind of food each enzyme acts on.

Enzyme	Body Source	Food Acted On
Ptyalin		
Amylase		
Lactase		
Pepsin		
Lipase		
Trypsin		
Maltase		
Rennin		
Erepsin		
Sucrase		

Activity   **Families of Chemical Compounds**   CHAPTER
                                                                                      3

Science Concentration

Science Concentration can be played with one or two friends or classmates. Before beginning the game, cut out the word cards on the next three pages. After you have done this, shuffle all the cards and place them singly face down on a table or desk.

To start the game, a player turns over two cards, one at a time. If these two cards contain a vocabulary word and its correct definition or example, that player scores one point and the cards are removed from the playing area. If the cards do not match, they are returned to their original positions. Another player then turns over two cards, one at a time. Again, cards are removed from the playing area only if a word match is made. Players continue to take turns until all the words have been matched. The player with the most points (matches) wins the game.

Mixture	**Milk, toothpaste, mayonnaise, or whipped cream**
Physical properties	**Brass, stainless steel, or bronze**
Heterogeneous mixture	**A substance that dissolves in water**

Chemistry of Matter   O ■ 71

Suspension	A substance that does not dissolve in water
Homogeneous mixture	Two or more pure substances that are mixed but not chemically combined
Alloy	The substance that is dissolved in a solution
Colloid	The substance in a solution that does the dissolving
Solution	Basis for separating mixtures
Solute	Mixture in which no two parts are identical

Solvent	**A mixture of oil and vinegar**
Insoluble	**A "well-mixed" mixture in which different parts seem to be identical**
Soluble	**The "best mixed" mixture in which the particles are dissolved in one another**

Activity _____ CHAPTER

Families of Chemical Compounds **3**

Solutions Everywhere

Using the three general phases of matter—solid, liquid, and gas—you can come up with nine different types of solutions. Listed below are the nine possible combinations of solutes and solvents. Under each combination, give an example of this type of solution. Then in the next two columns, describe the specific solute and solvent that make up your solution. Look around you for examples; there are solutions everywhere!

Type of Solution	Specific Solute	Specific Solvent
Gas in Gas _____		
Gas in Liquid _____		
Gas in Solid _____		
Liquid in Gas _____		
Liquid in Liquid _____		
Liquid in Solid _____		
Solid in Gas _____		
Solid in Liquid _____		
Solid in Solid _____		

Chemistry of Matter O ■ 75

Activity

Families of Chemical Compounds

CHAPTER
3

Solutions and Temperature

Some solutes dissolve better when the solution is heated. Other solutes dissolve better when the solution is cooled. The following exercise illustrates the importance of temperature in dissolving substances in a solution.

Listed below are the approximate amounts of two different solutes that will dissolve in 100 g of water at several different temperatures. To more easily compare how temperature affects the solubility of these two solutes, prepare a graph of these data. The horizontal, or X, axis of your graph should be Temperature. The vertical, or Y, axis should be Grams of solute dissolved per 100 g of water. Put both solutions on the same graph, but use a different color for each substance. Be sure to make the intervals between values of a size that uses most of the graph paper.

Temperature	Sucrose $C_{12}H_{22}O_{11}$	Cerium Sulfate $Ce_2(SO_4)_3$
(degrees Celsius)	(approximate grams dissolved per 100 g H_2O)	(approximate grams dissolved per 100 g H_2O)
0	180	21
5	185	18
10	190	16
15	197	14
20	204	12
25	209	10
30	220	9
35	230	8
40	239	7

After preparing your graph, use the solubility lines to answer the following questions.

1. At 33°C, which solution would have the most solute dissolved? _____

Which substance is most soluble? _____

2. Not all solutes dissolve when they are warmed. In addition to cerium sulfate, can you

think of two other examples? _____

3. If you wanted more sucrose to dissolve in water, like sugar dissolving in tea, would

you heat the solution or cool it? _____

If you wanted more cerium sulfate to dissolve in water, would you heat the solution

or cool it? _____

4. What effect do you think increasing the pressure directly over the solutions would have on the solubility of each solute? _____

5. Between 5°C and 10°C, which solute shows the greatest change in the amount that can be dissolved in 100 g of H_2O? _____

6. Over the range of temperatures given, which solute shows the greatest change in solubility?

_____

7. Here is a question that requires some thought. Solutes dissolving in a solvent will usually either release energy or gain energy and thus cause the solution to warm or cool to your touch. From the data in this exercise, can you tell which solution process would cause a slight cooling? A slight warming? Explain your answer.

_____

_____

_____

_____

_____

ctivity

Naming Hydrocarbons

Hydrocarbons are compounds made up of carbon and hydrogen. Hydrocarbons called alkanes are the simplest hydrocarbons. These compounds are named by using a prefix that tells the number of carbon atoms they contain and the root *-ane*.

Using the chart below, name each of the alkanes that are shown.

Prefix	Number of Carbon Atoms
meth-	1
eth-	2
prop-	3
but-	4
pent-	5
hex-	6
hept-	7
oct-	8
non-	9
dec-	10

1. _____

2. _____

3. _____

4. _____

5. _____

6. _____

7. _____

_____ *Laboratory Investigation* _____

CHAPTER 3 ■ Families of Chemical Compounds

Acids, Bases, and Salts

Problem

What are some properties of acids and bases? What happens when acids react with bases?

Materials *(per group)*

safety goggles	test-tube rack
beaker	red and blue
solutions of	litmus paper
H_2SO_4, HCl,	stirring rod
HNO_3	medicine dropper
solutions of	phenolphthalein
KOH, NaOH,	evaporating dish
$Ca(OH)_2$	
6 medium-sized	
test tubes	

Procedure 🔥 ⚗ 👁

A. Acids

1. Put on your safety goggles. Over a sink, pour about 5 mL of each acid into separate test tubes. **CAUTION:** *Handle acids with extrme care. They can burn the skin.* Place the test tubes in the rack. Test the effect of each acid on litmus paper by dipping a stirring rod into the acid and then touching the rod to the litmus paper. Test each acid with both red and blue litmus paper. *Be sure to clean the rod between uses.* Record your observations.

2. Add 1 drop of phenolphthalein to each test tube. Record your observations.

B. Bases

1. Over a sink, pour about 5 mL of each base into separate test tubes. **CAUTION:** *Handle bases with extreme care.* Place the test tubes in the rack. Test the contents of each tube with red and blue litmus paper. Record your observations.

2. Add 1 drop of phenolphthalein to each test tube. Record your observations.

3. Place 5 mL of sodium hydroxide solution in a small beaker and add 2 drops of phenolphthalein. Record the solution's color.

4. While slowly stirring, carefully add a few drops of hydrochloric acid until the mixture changes color. Record the color change. This point is known as the indicator endpoint. Test with red and blue litmus paper. Record your observations.

5. Carefully pour some of the mixture into a porcelain evaporating dish. Let the mixture evaporate until it is dry.

Observations

1. What color do acids turn litmus paper? Phenolphthalein?

_____

2. What color do bases turn litmus paper? Phenolphthalein?

_____

3. What happens to the color of the sodium hydroxide-phenolphthalein solution when hydrochloric acid is added?

_____

4. Does the substance formed by the reaction of sodium hydroxide with hydrochloric acid affect litmus paper?

_____

5. Describe the appearance of the substance that remains after evaporation.

_____

Analysis and Conclusions

1. What are some properties of acids? Of bases?

_____

_____

_____

2. What type of substance is formed when an acid reacts with a base? What is the name of this reaction? What is the other product? Why does this substance have no effect on litmus paper?

_____

_____

_____

3. What is meant by an indicator's endpoint?

_____

_____

4. On Your Own   Write a balanced equation for the reaction between sodium hydroxide and hydrochloric acid.

_____

Answer Key

Chapter Discovery: What's in a Flavor?

Observations: apricot; pineapple; wintergreen; grape **Critical Thinking and Application** **1.** ethyl alcohol and butyric acid **2.** isoamyl alcohol and butyric acid **3.** methyl alcohol and salicylic acid **4.** ethyl alcohol and decanoic acid **5.** ethyl butyrate, isoamyl butyrate, methyl salicylate; ethyl decanoate

Activity: Names and Uses of Salts

Sodium bicarbonate baking soda; soothes upset stomach and skin irritations, used as a leavening agent **Potassium nitrate, sodium nitrate** saltpeter; used in the manufacture and production of fertilizers, matches, and explosives **Sodium carbonate** sal soda, washing soda; used to neutralize acids, disinfect, clean and soften water; used in the manufacture of soap and glass **Silver nitrate** none; used as an antiseptic and in making film **Copper sulfate** blue vitriol; used in the dying and calico printing industries **Potassium bromide** none; used in medicines

Discovery Activity: Acid, Base, or Neutral Substance?

1a. acid **b.** below 7 **2a.** acid **b.** below 7 **3a.** base **b.** above 7 **4a.** neutral **b.** 7 **5a.** base **b.** above 7 **6a.** neutral **b.** 7

Discovery Activity: Dissolving Household Substances

Answers will vary depending on the substances they test. Students can use a teaspoon to measure the amount of substance used.

Discovery Activity: Acid, Base, or Salt?

H_3PO_4 acid, red, clear, below 7 $Ca(OH)_2$ base, blue, red, above 7 $CaSO_4$ salt, no effect, clear, about 7 H_2CO_3 acid, red, clear, below 7 Na_3PO_4 salt, no effect, clear, about 7 H_2S acid, red, clear, below 7 H_2SO_3 acid, red, clear, below 7 $Mg(OH)_2$ base, blue, red, above 7 NH_4Cl salt, no effect, clear, about 7 $(NH_4)_2SO_4$ salt, no effect, clear, about 7

Activity: Digestion

After a period of time the crackers begin to taste sweet. An enzyme in your saliva, ptyalin, has changed the starch present in the cracker into sugar.

Ptyalin salivary glands, starches **Amylase** pancreas, starches **Lactase** small intestine, lactose (milk sugar) **Pepsin** stomach, proteins **Lipase** pancreas, fats **Trypsin** pancreas, proteins **Maltase** small intestine, maltose (malt sugar) **Rennin** stomach, proteins **Erepsin** small intestine, peptides **Sucrase** small intestine, compound sugars

Activity: Science Concentration

Mixture Two or more pure substances that are mixed together but not chemically combined **Physical properties** Basis for separating mixtures **Heterogeneous mixture** Mixture in which no two parts are identical **Suspension** A mixture of oil and vinegar **Homogeneous mixture** A "well-mixed" mixture, in which different parts seem to be identical **Alloy** Brass, stainless steel, or bronze **Colloid** Milk, toothpaste, mayonnaise, or whipped cream **Solution** The "best-mixed" mixture, in which the particles are dissolved in one another **Solute** The substance that is dissolved in a solution **Solvent** The substance in a solution that does the dissolving **Insoluble** A substance that does not dissolve in water **Soluble** A substance that dissolves in water

Discovery Activity: Solutions Everywhere

Answers will vary. Examples follow. **Row 1** air, oxygen, nitrogen **Row 2** ammonia cleaners, ammonia, water **Row 3** catalytic converters in cars, hydrocarbons, platinum **Row 4** no actual solutions, but fog comes close, water, air **Row 5** gasoline, octane (C_8H_{18}), heptane (C_7H_{16}) **Row 6** mercury batteries, mercury, zinc **Row 7** no actual solutions, but smoke comes close, carbon, air **Row 8** iced tea, sugar, water **Row 9** steel, carbon, iron

Problem-Solving Activity: Solutions and Temperature

1. sucrose, sucrose **2.** O_2 and CO_2 **3.** heat the solution, cool the solution **4.** No effect. Pressure affects the solubility of gases but has no appreciable effect on the solubility of solids. **5.** sucrose: 5 g/100 g of H_2O **6.** sucrose: 59 g **7. slight cooling:** sucrose. Energy is used to dissolve sucrose, as can be seen from the graph. If you were holding the glass of H_2O in which the sucrose was being dissolved, the solution would absorb energy from your hand (and the air). Because your hand loses energy (a very slight amount), it would feel cool. **slight warming:** cerium sulfate. Cerium sulfate dissolves better as energy is taken away from the solution. The energy goes into the immediate surroundings, and your hand would feel slightly warm.

Activity: Naming Hydrocarbons

1. butane **2.** methane **3.** propane **4.** ethane **5.** hexane **6.** pentane **7.** octane

Laboratory Investigation: Acids, Bases, and Salts

Observations **1.** Red; colorless **2.** Blue; bright pink **3.** The solution is colorless **4.** No **5.** White powder **Analysis and Conclusions** **1.** Acids turn litmus paper red and phenolphthalein colorless. Acids react with bases to form neutral substances. Bases turn litmus paper blue and phenolphthalein bright pink **2.** A salt; neutralization; water; both the salt and the water are neutral. **3.** An indicator's endpoint is the point at which a change in color indicates a neutral substance or the absence of either hydrogen or hydroxide ions in solution **4.** NaOH + HCl = NaCl + H_2O

Contents

Chapter Discovery

Chapter Activities

Laboratory Investigation Worksheet

   (**Note:** *This investigation is found on page O106 of the student textbook.*)

*Appropriate for cooperative learning

— C —————————————————————————————————— CHAPTER

Chapter Discovery   **Petrochemical Technology**   **4**

Making Aspirin

Background Information

Ethylene (C_2H_4) is a hydrocarbon that can be obtained from the fractional distillation of petroleum. When ethylene undergoes certain chemical reactions, the organic acid commonly known as acetic acid is formed. It is from acetic acid that the familiar medicine aspirin is produced. In this activity, you will use the anhydrid form (meaning that water molecules have been removed) of acetic acid to make your own aspirin.

Materials

salicylic acid	filter paper
acetic anhydride	medicine dropper
concentrated sulfuric acid	large beaker
water	graduated cylinder
ice	paper towel
bucket	heat source
balance scale	ring stand
thermometer	timer
Erlenmeyer flask	several commercial aspirin tablets
funnel	mortar and pestle

Procedure ⚗ 🔥 ☢ ✊ 👁

1. Put on your safety goggles and a laboratory apron.

2. Measure 3 g of salicylic acid.

3. Place the salicylic acid in the Erlenmeyer flask. Slowly add 6 mL of acetic anhydride. **CAUTION:** *Pour the acetic anhydride carefully. Be sure to observe all safety rules for working with chemicals.*

4. Add 5 drops of sulfuric acid to the flask. **CAUTION:** *Be very careful when working with acid. Do not allow any acid to splash onto your skin or clothing.* Gently swirl the flask to mix the chemicals.

5. Fill a large beaker about half full with water. Place the flask in the beaker so that the part of the flask containing the chemicals is submerged. Position a thermometer so you can measure the temperature of the water.

6. Heat the beaker until the temperature of the water is 80°C. Continue heating the beaker for 10 minutes. Do not allow the temperature of the water to go above 90°C.

7. Remove the flask from the beaker and allow to cool to room temperature. Then fill a bucket with ice and place the flask in the ice. Add 40 mL of water to the flask. What do you see happening?

_____

_____

8. Place a sheet of filter paper in a funnel. Use a ring stand to support the funnel. Place a beaker under the opening of the funnel.

9. Pour the contents of the flask into the funnel. Discard the liquid that flows into the beaker. Describe the substance that remains in the funnel.

_____

_____

_____

10. Wash the substance with a small amount of cold water. (If the substance is brown, wash again and repeat step 9.)

11. Gently blot the substance between layers of paper toweling. Leave the substance to dry in the open air. After several minutes, place a commercial aspirin next to the substance you have made. Compare their appearances.

_____

_____

_____

12. Crush the commercial aspirin tablet with a mortar and pestle. Now how does the aspirin compare with the substance you have made?

_____

_____

_____

Critical Thinking and Application

1. From what substances is aspirin made?

_____

_____

2. What are some physical properties of aspirin?

_____

_____

3. Although aspirin is one of the most useful drugs known, it is one of the least expensive. Can you think of a reason for this?

_____

_____

Activity

A World of Synthetic Polymers

The picture below contains 30 objects that are synthetic polymers. These objects range from polyvinyl plastics and Formica to polyesters, vinyls, and synthetic rubber. How many synthetic polymers can you find in the picture? Study the picture carefully. Then put a number next to each synthetic polymer you find.

Activity

Constructing a Nylon Polymer Molecule

The molecule in the diagram below is a monomer of nylon. A monomer is a single unit of atoms that is repeated over and over again to form a chained molecule called a polymer. In this activity you will construct a model of the nylon polymer. All you need to do this is paper, scissors, and a stapler. Follow the steps listed below and have fun while making nylon.

1. Cut out the rectangular box along the lines indicated in the diagram below.

2. Cut three more rectangular boxes exactly like this one from another piece of paper.

3. Tape the four pieces (the original and three copies) end-to-end. You should have one long, narrow strip of paper.

4. Fold the paper accordian style (like a fan) at the joined ends. Make sure your original is on top and the outline of the atoms is showing.

5. Staple the four folded sections together at each corner. **Note:** *Do not staple over the outline of the atoms.*

6. Now cut along the outline of the atoms. Do this slowly and carefully.

7. Unfold your masterpiece. You now have a nylon polymer. Label each atom with the appropriate symbol.

cut along this line

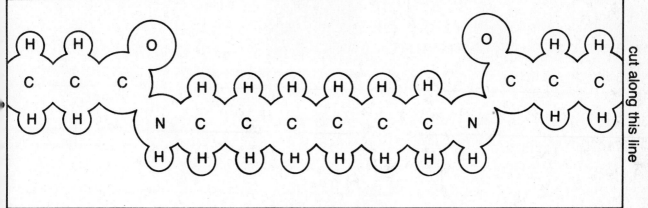

cut along this line

Activity

Bakelite

Bakelite is the name of the first synthetic polymer. Its development is an interesting story. Using books and other reference materials in the library, find out about the discovery of this polymer. Use the space below to outline the main ideas of your research. Be sure to include the following information: how, when, and by whom Bakelite was discovered, and the characteristics and commercial uses of Bakelite.

_____

_____

_____

_____

_____

_____

_____

_____

_____

_____

_____

_____

_____

_____

_____

_____

_____

_____

_____

_____

_____

_____

_____

Activity

CHAPTER
4

A Model Oil Reservoir

Geologists use different clues to hunt down likely sources of oil deep within the Earth. These include fossils, the depth and kind of rock layers in the Earth, the presence or absence of faults, and the way rock layers are folded. Using books and other reference materials in the library, find out about the processes involved in oil exploration.

1. Make a clay model of a cross section of the Earth in which an oil reservoir might be located. Use different colors of clay to show the various layers of the Earth. In the space below, draw a diagram of your model.

2. Describe the "clues" that indicate that this location is a good place to drill for oil.

_____

_____

_____

_____

_____

Fractional Distillation of Petroleum

The figure is a diagram of a fractionating tower. Use the table to determine which substance will condense at each level. Write the name of each substance in the space provided at the correct level of the fractionating tower.

Substance	Condensing Temperature (°C)
Asphalt	Above 400
Jet fuel	180–275
Industrial fuel oil	325–350
Heating fuels	190–275
Gasoline	30–175
Diesel fuel	275–325
Kerosene	175–275
Lubricating oils	About 350

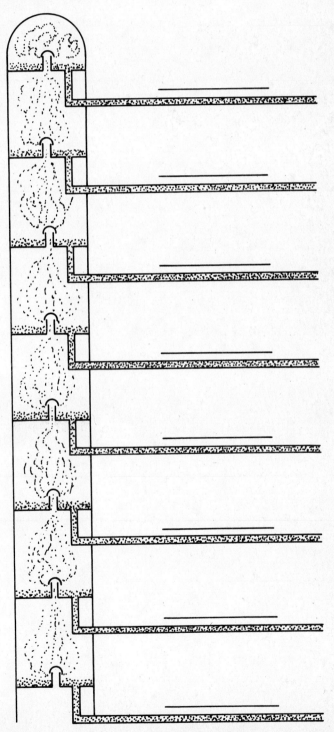

_____ *Laboratory Investigation* _____

CHAPTER 4 ■ Petrochemical Technology

Comparing Natural and Synthetic Polymers

Problem
How do natural and synthetic polymers compare in strength, absorbency, and resistance to chemical damage?

Materials (*per group*)
3 samples of natural polymer cloth:
    wool, cotton, linen
3 samples of synthetic polymer cloth:
    polyester, nylon, acetate
12 Styrofoam cups
mild acid (lime or lemon juice or vinegar)

marking pen	liquid bleach
metric ruler	medicine dropper
scissors	oil
rubber gloves	paper towel

Procedure 🔥 ⚗ 👁 ▰

1. Record the color of each cloth.

2. Label 6 Styrofoam cups with the names of the 6 cloth samples. Also write the word Bleach on each cup.

3. Cut a 2-square-cm piece from each cloth. Put each piece in its cup.

4. Wearing rubber gloves, carefully pour a small amount of bleach into each cup.

5. Label the 6 remaining cups with the names of the 6 cloth samples and the word Acid. Then pour a small amount of the mild acid into each and repeat step 3.

6. Set the cups aside for 24 hours. Meanwhile, proceed with steps 7 through 9.

7. Using the remaining samples of cloth, attempt to tear each.

8. Place a drop of water on each material. Note whether the water forms beads or is absorbed. If the water is absorbed, record the rate of absorption.

9. Repeat step 8 using a drop of oil.

10. After 24 hours, wearing rubber gloves, carefully pour the liquids in the cups into the sink or a container provided by your teacher. Dry the samples with a paper towel.

11. Record any color changes.

Observations

1. Which material held its color best in bleach? In acid?

_____

_____

2. Which materials were least resistant to chemical damage by bleach or mild acid?

_____

_____

3. Which material has the strongest fiber or is hardest to tear?

_____

_____

4. Which materials are water repellent?

_____

_____

Analysis and Conclusions

1. Compare the natural and synthetic polymers' strength, absorbency, and resistance to chemical change.

_____

_____

_____

_____

2. Which material would you use to manufacture a laboratory coat? A farmer's overalls? A raincoat? An auto mechanic's shirt?

_____

_____

_____

_____

3. **On Your Own** Confirm your results with additional samples of natural and synthetic polymers. What additional tests can you add for comparison?

_____

_____

_____

Answer Key

Chapter Discovery: Making Aspirin

7. White crystals form.   **9.** White crystalline solid   **11.** Appearances should be similar except that the commercial aspirin has been pressed into a tablet.   **12.** They look the same.   **Critical Thinking and Application 1.** acetic acid, salicylic acid, and a small amount of sulfuric acid   **2.** white color; made of crystals   **3.** Students should recognize the ease of aspirin's preparation. Also students should realize that aspirin is made from ingredients that are easy to obtain.

Discovery Activity: A World of Synthetic Polymers

There are 30 polymers in the picture. Some of them include pot handles, light coverings, floor tiles, blinds, jacket, oven door, bowls, utensils, pot lids, canisters, and a coffee pot.

Activity: Constructing a Nylon Polymer Molecule

Check models to make sure that students have followed directions.

Activity: Bakelite

Bakelite was invented in 1909 by Leo H. Baekeland, an American chemist. The plastic was used to make electrical insulating materials, pipe stems, and radio cabinets. Bakelite is a compound made from phenol and formaldehyde. It was the first synthetic resin made.

Activity: A Model Oil Reservoir

Check models for accuracy. Scientists use explosives and other special devices to detect movements under the Earth that indicate an oil deposit. Students can report on these and other "clues."

Problem-Solving Activity: Fractional Distillation of Petroleum

1. gasoline   **2.** kerosene   **3.** jet fuel **4.** heating fuels   **5.** diesel fuel   **6.** industrial fuel oil   **7.** lubricating oils   **8.** asphalt

Laboratory Investigation: Comparing Natural and Synthetic Polymers

Observations   1. Wool, wool.   **2.** Cotton, linen.   **3.** Polyester.   **4.** Polyester, nylon. **Analysis and Conclusions   1.** The synthetic polymer polyester was the strongest in terms of being hardest to tear. The synthetic polyester absorbed the least amount of water, whereas the synthetic polymers polyester and nylon absorbed the least amount of oil. The natural polymer wool was the most resistant to chemical wear from bleach and acid.   **2.** Lab coat: cotton, linen; farmer's overalls: polyester; raincoat: polyester, nylon; auto mechanic's shirt: polyester, nylon. **3.** Students may suggest testing the fabric samples for fire resistance or stain resistance.

Contents

CHAPTER 5 ■ Radioactive Elements

*Appropriate for cooperative learning

Chapter Discovery

Modeling a Decay Series

Background Information

Atoms of certain elements are unstable. This means that they cannot remain in their original form. To become stable, they may undergo a process called radioactive decay. Radioactive decay is the spontaneous breakdown of an unstable atomic nucleus. During radioactive decay, three types of radiation may be emitted from the nucleus. The three types of radiation are called alpha particles, beta particles, and gamma rays.

In some cases, even after the atom undergoes radioactive decay, the nucleus is still unstable. When this happens, the nucleus again decays. This process continues until a stable nucleus is formed. This pattern of radioactive decay is called a decay series. In this activity, you will model part of the decay series for uranium-238.

Materials:

about 250 small dried beans
dark-colored food dye or spray paint
6 large paper plates
yellow construction paper
ruler
scissors
glue
marking pen

Procedure ▆

1. Color about 100 small dried beans. You can do this either by soaking the beans overnight in water that has been colored with dark food dye or by spraying the beans with paint. The colored beans represent protons. Collect another 150 beans to represent neutrons.

2. Cut five squares of yellow construction paper, making each square about 6 cm × 6 cm.

3. Line up six paper plates in a row, leaving about 15 cm between plates.

4. Write the symbol U-234 and 92-P/142-N on the first plate. Place 92 colored beans and 142 uncolored beans on the plate.

5. Take 2 colored beans and 2 uncolored beans off the plate and place them on a yellow square. This square represents an alpha particle. Place this yellow square between the first and second plate.

6. Write Th-230 and 90-P/140-N on the second plate. Take the beans from the first plate and place them on the second plate. Leave the first plate in its original position.

7. Take 2 colored beans and 2 uncolored beans off the second plate and place them on another yellow square. Place this yellow square between the second and third plates.

8. Write Ra-226 and 88-P/138-N on the third plate. Take the beans from the second plate and place them on the third plate. Leave the second plate in its position.

9. Remove 2 colored beans and 2 uncolored beans from the third plate and place them on another yellow square. Place the square between the third and fourth plates.

10. Write Rn-222 and 86-P/136-N on the fourth plate. Move the beans from the third plate onto the fourth plate.

11. Remove 2 colored beans and 2 uncolored beans from the fourth plate. Glue them to another yellow square. Place this square between the fourth and fifth plates.

12. On the fifth plate, write Po-218. Then write 84-P and 134-N. Move the beans from the fourth plate onto the fifth plate.

13. From the fifth plate, take 2 colored beans and 2 uncolored beans and glue them to the last yellow square. Place this square between the fifth and sixth plates.

14. On the sixth plate, write Pb-214; then write 82-P and 132-N. Move the beans from the fifth plate onto the sixth plate. Try to arrange the beans so you can still see the numbers you have written.

Critical Thinking and Application

1. If each of the paper plates represents an atomic nucleus, how many different nuclei were involved in this series? Write the names of the elements in order with the mass number of each.

_____

_____

_____

2. How many alpha particles were emitted each time an element changed? How many all together?

_____

_____

_____

3. What subatomic particles make up an alpha particle?

_____

_____

4. How did the emission of an alpha particle affect the atomic number of a nucleus? The mass number?

_____

_____

5. Lead-214 is a radioactive isotope of lead. What do you think will happen to this nucleus? Why?

_____

_____

Activity

Radioactive and Nonradioactive Elements at a Glance

1. On a large sheet of paper or posterboard, make a copy of the periodic table found on text pages 154 and 155.

2. Use different colors to show which elements are radioactive, which are nonradioactive, and which are synthetic.

3. Describe the pattern you see. Which elements are unstable? Stable?

_____

_____

_____

_____

_____

_____

_____

_____

_____

_____

_____

_____

_____

_____

_____

_____

_____

_____

_____

_____

_____

_____

Activity

Alpha, Beta, and Gamma Radiation in a Magnetic Field

Charged particles are deflected by a magnetic field. Positive charges are deflected in the opposite direction of negative charges. More massive particles will be deflected less than lighter particles. However, an uncharged particle will not be deflected by a magnetic field. There are three types of radioactive energy: alpha, beta, and gamma radiation. The alpha and beta particles are oppositely charged. The alpha particle is much heavier than the beta particle. The gamma particle has no charge. Identify the paths of the alpha, beta, and gamma radiation in the following diagram. Write the correct name in the circles provided.

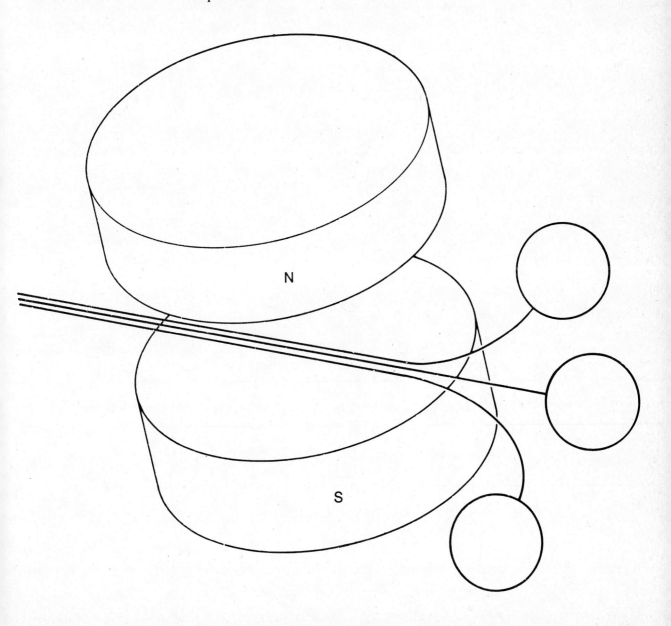

Nuclear Fission and Nuclear Fusion

Use the following titles and terms to correctly identify and label the diagrams below.

NUCLEAR FISSION
Energy
Uranium-235 nucleus
2 or 3 neutrons
Krypton-92 nucleus
Neutron
Barium-141 nucleus

NUCLEAR FUSION	
$^2_1 H$	$^4_2 He$
$^2_1 H$	$^1_1 H$
$^1_1 H$	$^1_1 H$
$^1_1 H$	Energy

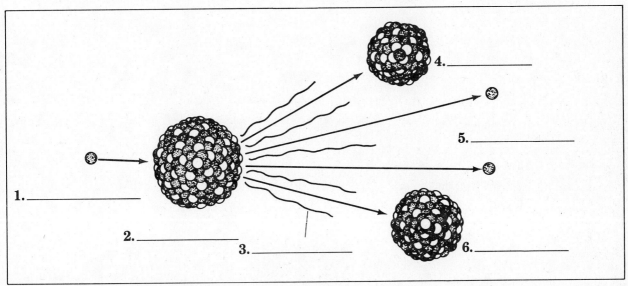

1. _____

2. _____

3. _____

4. _____

5. _____

6. _____

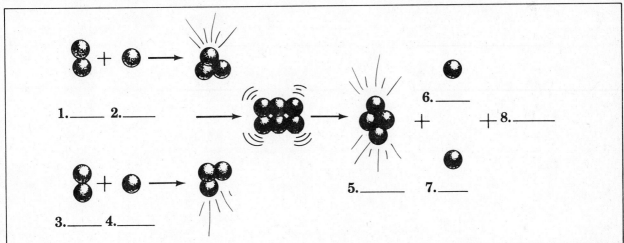

1. ____ 2. ____

3. ____ 4. ____

5. ____ 6. ____ 7. ____ 8. ____

Decay Series of Uranium-238

When a radioactive element decays by emitting an alpha (α) or a beta (β) particle, a new element is formed. This process is called transmutation. An alpha particle contains two protons and has an atomic mass of four. During alpha decay, an atom loses two protons and four atomic mass units. Since the atomic number of an element is the number of protons, the atomic number of the atom is decreased by two. The mass number is decreased by four.

In nuclear science an element is identified in the following way.

$$\begin{matrix} A \\ E \\ Z \end{matrix} \quad \begin{matrix} \text{Where A = atomic mass} \\ \text{E = element} \\ \text{Z = atomic number} \end{matrix}$$

Uranium-238 is $^{238}_{92}U$. Uranium-238 has an atomic mass of 238 and an atomic number of 92. When uranium-238 decays, it emits an alpha particle and becomes element number 90 (92 − 2), which is thorium. The mass of the thorium atom is 234 (238 − 4). This decay can be represented in the following way.

$$^{238}_{92}U \xrightarrow{\;\alpha\;} {}^{234}_{90}Th$$

The beta particle is a high-speed electron that is emitted by the nucleus. When a neutron in the nucleus decays, an electron is released and a proton is left behind (1 neutron = 1 proton + 1 electron). The atomic number, therefore, increases by one proton. Since the mass of an electron is so small, the atomic mass remains the same. So beta decay increases the atomic.number by one, but does not change the atomic mass. Thorium-234 emits a beta particle and transmutes to element 91, which is protactinium.

$$^{234}_{90}Th \xrightarrow{\;\beta\;} {}^{234}_{91}Pa$$

Notice that the atomic mass, 234, does not change.

In the following U-238 decay series, fill in the correct number or element in each of the circles. You will need a periodic table of the elements to identify an element from its atomic number.

$$^{238}_{92}U \xrightarrow{\alpha} {}^{234}Th \xrightarrow{\beta} {}_{91}Pa \xrightarrow{\;} {}^{234}_{92}U \xrightarrow{\;} {}^{230}_{90}Th \xrightarrow{\alpha} {}^{226}_{88} \bigcirc \xrightarrow{\alpha} {}^{222}Rn \xrightarrow{\alpha}$$

$$_{84}Po \xrightarrow{\alpha} {}_{82}Pb \xrightarrow{\beta} {}^{214}Bi \xrightarrow{\;} {}^{214}_{84}Po \xrightarrow{\alpha} {}^{210}Pb \xrightarrow{\;} {}^{210}_{83}Bi \xrightarrow{\beta} {}^{210}_{84} \bigcirc \xrightarrow{\;} {}^{206}_{82}Pb$$

Activity _____ CHAPTER

Radioactive Elements **5**

Atomic Arithmetic

Fill in the number that answers each statement and do the arithmetic indicated. Your final answer will be the value for the half-life of carbon-14. The dotted lines will contain your answers to each intermediate step.

1. The mass number of the uranium isotope most commonly used as a nuclear fuel _____

2. The number of grams of a 100-g sample of rhodium-106 that remains after 60 seconds, if rhodium's half-life is 30 seconds × _____

    - - - - - - -

3. The atomic number of a thorium atom that forms when $^{92}_{238}U$ undergoes alpha decay − _____

    - - - - - - -

4. The number of protons in the nucleus of a uranium atom − _____

    - - - - - - -

5. The increase in the atomic number of an element as a result of releasing a beta particle + _____

    - - - - - - -

6. The number of quarks that make up a proton ÷ _____

    - - - - - - -

7. The atomic number of an atom having 3 protons × _____

    - - - - - - -

8. The number of protons plus neutrons in an alpha particle − _____

    - - - - - - -

9. The mass number of the isotope of barium that results from the fission of a uranium-235 nucleus − _____

    - - - - - - -

10. The mass number of a fluorine atom having 9 protons and 10 neutrons + _____

    - - - - - - -

The half-life of carbon-14 is _____ years.

Nuclear Equations

Complete each nuclear equation by filling in the blank space.

1. $^{214}_{84}\text{Po} \rightarrow {}^{210}_{82}\text{Pb} + $ _____

2. $^{222}_{86}\text{Rn} \rightarrow $ _____ $+ {}^{4}_{2}\text{He}$

3. $^{230}_{90}\text{Th} \rightarrow {}^{226}_{88}\text{Ra} + $ _____

4. $^{214}_{82}\text{Pb} \rightarrow $ _____ $+ {}^{0}_{-1}\text{e}$

5. $^{226}_{88}\text{Ra} \rightarrow $ _____ $+ {}^{4}_{2}\text{He}$

6. $^{239}_{93}\text{Np} \rightarrow $ _____ $+ {}^{0}_{-1}\text{e}$

7. _____ $\rightarrow {}^{234}_{90}\text{Th} + {}^{4}_{2}\text{He}$

8. $^{234}_{92}\text{U} \rightarrow {}^{234}_{93}\text{Np} + $ _____

9. $^{206}_{82}\text{Pb} + {}^{4}_{2}\text{He} \rightarrow $ _____

10. $^{254}_{91}\text{Pa} \rightarrow $ _____ $+ {}^{0}_{-1}\text{e}$

11. $^{226}_{88}\text{Ra} \rightarrow $ _____ $+ {}^{4}_{2}\text{He} + \text{gamma rays}$

12. $^{214}_{82}\text{Pb} \rightarrow $ _____ $+ {}^{0}_{-1}\text{e} + \text{gamma rays}$

Laboratory Investigation

CHAPTER 5 ■ Radioactive Elements

The Half-Life of a Sugar Cube

Problem
How can the half-life of a large sample of sugar cubes be determined?

Materials *(per group)*
250 sugar cubes   large bowl
food coloring   medicine dropper

Procedure 🔺
1. Place a small drop of food coloring on one side of each sugar cube.
2. Put all the sugar cubes in a bowl. Then gently spill them out on the table. Move any cubes that are on top of other cubes.
3. Remove all the sugar cubes that have the colored side facing up. If you have room on the table, arrange in a vertical column the sugar cubes that you removed. Put the rest of the cubes back in the bowl.
4. Repeat step 3 several more times until five or fewer sugar cubes remain.
5. On a chart similar to the one shown, record the number of tosses (times you spilled the sugar cubes), the number of sugar cubes removed each time, and the number of sugar cubes remaining. For example, suppose after the first toss you removed 40 sugar cubes. The number of tosses would be 1, the number of cubes removed would be 40, and the number of cubes remaining would be 210 (250−40).

Observations
1. Make a full-page graph of tosses versus cubes remaining. Place the number of tosses on the X (horizontal) axis and the number of cubes remaining on the Y (vertical) axis. Start at zero tosses with all 250 cubes remaining.
2. Determine the half-life of the decaying sugar cubes in the following way. Find the point on the graph that corresponds to one half of the original sugar cubes (125). Move vertically down from this point until you reach the horizontal axis. Your answer will be the number of tosses.

Tosses	Sugar Cubes Removed	Sugar Cubes Remaining
0	0	250
1	40	210
2		
3		

Analysis and Conclusions

1. What is the shape of your graph?

   _____

2. How many tosses are required to remove one half of the sugar cubes?

   _____

3. How many tosses are required to remove one fourth of the sugar cubes?

   _____

4. Assuming tosses are equal to years, what is the half-life of the sugar cubes?

   _____

5. Using your answer to question 4, how many sugar cubes should remain after 8 years? After 12 years? Do these numbers agree with your observations?

   _____

   _____

   _____

6. What factor(s) could account for the differences in your observed results and those calculated?

   _____

   _____

   _____

7. **On Your Own** Repeat the experiment with a larger number of sugar cubes. Predict whether the determined half-life will be different. Is it?

   _____

   _____

Answer Key

Chapter Discovery: Modeling a Decay Series

1. six; Uranium-234; Thorium-230; Radium-226; Radon-222; Polonium-218; Lead-214 **2.** one per change; five all together **3.** 2 protons and 2 neutrons **4.** The emission of an alpha particle reduces the atomic number by 2 and the mass number by 4. **5.** The nucleus will break down again.

Activity: Radioactive and Nonradioactive Elements at a Glance

2. Elements that have an atomic number greater than 84 (84–109) are radioactive. Additionally, atomic numbers 43 and 61 are radioactive. All other elements with an atomic number of 83 or less are not radioactive. Synthetic elements have atomic numbers 93 or greater. **3.** In general, larger atoms are radioactive, smaller atoms are not. Radioactive elements are unstable. Nonradioactive elements are stable.

Activity: Alpha, Beta, and Gamma Radiation in a Magnetic Field

top circle, alpha; middle circle, gamma; bottom circle, beta

Activity: Nuclear Fission and Nuclear Fusion

Nuclear fission 1. neutron **2.** U-235 nucleus **3.** energy **4.** krypton-92 nucleus **5.** 2 or 3 neutrons **6.** barium 141 nucleus **Nuclear fusion 1.** 2 **2.** 1 **3.** 1 **4.** 1 **5.** $^4_2$He **6.** $^1_1$H **7.** $^1_1$H **8.** energy

Problem-Solving Activity: Decay Series of Uranium-238

90 234 β α α Ra 86 218 214 83 β 82 β α

Problem-Solving Activity: Atomic Arithmetic

1. 235 **2.** 25, 5875 **3.** 90, 5785 **4.** 92, 5693 **5.** 1, 5694 **6.** 3, 1896 **7.** 3, 5694 **8.** 4, 5690 **9.** 141, 5549 **10.** 19, 5568 5568 years

Activity: Nuclear Equations

1. $^4_2$He **2.** $^{218}_{84}$Po **3.** $^4_2$He **4.** $^{214}_{83}$Bi **5.** $^{222}_{86}$Rn **6.** $^{239}_{94}$Pu **7.** $^{238}_{92}$U **8.** $^0_{-1}$e **9.** $^{210}_{84}$Po **10.** $^{254}_{92}$U **11.** $^{222}_{86}$Rn **12.** $^{214}_{83}$Bi

Laboratory Investigation: The Half-Life of a Sugar Cube

Observations 1. Check students' graphs to see that each graph accurately reflects the data. **2.** Check students' determination of half-life to see that it reflects their graphed data. **Analysis and Conclusions 1.** Students may describe their graph using words or by drawing its shape. The ideal shape would display a hyperbolic curve in the first quadrant, asymptotic with respect to the X axis. **2.** Answers will vary, but approximately four tosses should remove one half of the cubes. **3.** Approximately 1 1/2 to 2 tosses should remove one fourth of the cubes. **4.** Four years. **5.** One fourth; one eighth. Students' data should be similar but probably not exact as the ideal figures given. **6.** Accept all logical answers. Half-life is a statistical measurement and cannot be considered exact using small sample sizes. As such, results from group to group, as well as from group to ideal, will vary. **7.** The half-life should not change considerably when more cubes are added. This can be inferred because the half-life of an element is the same regardless of how much of that element is considered.

Science Reading Skills

TO THE TEACHER

One of the primary goals of the *Prentice Hall Science* program is to help students acquire skills that will improve their level of achievement in science. Increasing awareness of the thinking processes associated with communicating ideas and reading content materials for maximum understanding are two skills students need in order to handle a more demanding science curriculum. Teaching reading skills to junior high school students at successive grade levels will help ensure the mastery of science objectives. A review of teaching patterns in secondary science courses shows a new emphasis on developing concept skills rather than on accumulating factual information. The material presented in this section of the Activity Book serves as a vehicle for the simultaneous teaching of science reading skills and science content.

The activities in this section are designed to help students develop specific science reading skills. The skills are organized into three general areas: comprehension skills, study skills, and vocabulary skills. The Science Gazette at the end of the textbook provides the content material for learning and practicing these reading skills. Each Science Gazette article has at least one corresponding science reading skill exercise.

C

Contents

Bucky Balls: New Adventures in Chemistry
Science Reading Skill: A Study Strategy

There is a difference between reading and studying. Studying involves more than reading; it includes sorting out information as you read. A good study guide follows the letter formula **PQRST**. Each letter refers to a step. The following is an explanation of each of the five steps.

Preview: Previewing means looking ahead to get an overview of the reading material. To preview, read the title and the first paragraph. Skim through the content. Pay attention to information that appears in **boldface** or in *italics*. Complete your preview by reading the last paragraph and any summary or study questions.

Question: Develop questions that will help you look for information to add to what you already know from your previewing. An easy way to do this is to change any headings or opening sentences of paragraphs into questions.

Read: Using the questions as guides, carefully but quickly read for important ideas. As you read, notice how these ideas relate to the topic you are studying.

Summarize: As you read, take notes by writing down key ideas in your own words. Write these notes in a notebook, not in your textbook.

Test: Compare your summary with the facts in the material you just read. Then ask yourself some questions to see if you remember the main points of what you have read. Make sure only important ideas have been included in your notes. Make a habit of reviewing your notes at different times. This will help you remember what you have learned.

To get a better idea of how the **PQRST** study plan works, apply the steps to the article on Bucky Balls. On your worksheet, you can list the information covered in your **P**review and **S**ummary. State four or five **Q**uestions after you have completed your preview. Time yourself when reading the article by writing down the time you began, the time you finish, and the total time you take to **R**ead the entire article. **T**est yourself by writing down four questions that your teacher might ask. Under each question, write the answer that you would supply if this were an actual test. This skill requires practice, but will prove to be one of the best gifts you can give yourself.

Worksheet for PQRST

Preview:

Question:

Read:

Summarize:

Test:

How Practical Is Flower Power?
Science Reading Skill: Drawing Conclusions

A conclusion is a final statement or decision that is formed when all the facts are present. A conclusion must be supported by evidence that is based on facts. The more facts there are to support a conclusion, the more reliable the conclusion is.

In studying science, the ability to draw conclusions is an important skill. Scientists make careful observations, collect data, and analyze information in order to form a conclusion. In a similar way, you can decide if a conclusion is reasonable from facts presented in the material. Here are some clues to help you.

1. See if the conclusion answers a question, solves a problem, or responds to an issue.

2. Look at the title or heading to find the topic of discussion.

3. Watch for indicator words that identify statements containing the conclusion. Examples of indicator words are *therefore, because, in short, this proves, it follows that.* Indicator words may be found at the beginning or at the end of the content material.

4. Make sure that enough facts are used to prove the conclusion.

Listed below are several statements containing conclusions. **Based on the information in this science article, write True or False in the space provided for each statement.**

1. _____ Burning of fuels is not a substitute for producing a new form of energy.

2. _____ Producing energy also requires an input of energy.

3. _____ Other forms of energy seem more reasonable to research than gopher plants.

4. _____ The corn plant is a practical source of gasohol.

5. _____ Eventually, Melvin Calvin will find a plant that directly produces fuel.

6. _____ Most of our energy will come from sources containing oil.

7. _____ Producing energy from plants is as inexpensive as it is practical.

8. _____ Determining the amount of hydrocarbons in gopher plants will affect the plants' possible use as a substitute fuel.

9. _____ Finding new forms of energy is a major problem.

10. _____ Farming fuel from plants is less expensive than other ways of getting energy.

Science Reading Skill: Effective Writing

In studying science, your skill in writing assignments, reports, and answers to test questions is very important. The purpose of written work is to communicate information. There are several basic rules you should follow when doing a writing exercise.

1. Have a good working knowledge of the subject you are writing about.

2. Use correct vocabulary and punctuation. This includes spelling, sentence structure, and paragraph form.

3. Use logical reasoning in developing ideas so that they apply to the information required.

4. Choose a topic sentence that states your purpose or identifies the main idea.

5. Always support your ideas by using factual information.

6. Reread what you have written to make sure that it is meaningful and represents your own thoughts.

A question appears at the end of this article. In two or three paragraphs, write your response. Apply the rules you have just learned to write an effective answer. Do some research so that you can support your answer. Use the space below for your written work.

The Right Stuff—Plastics
Science Reading Skill: Defining Technical Terms

Perhaps you have had difficulty with some of the scientific terms that you have come across in reading science material. One skill that can help you find the meaning of scientific terms is using context clues. In this technique, you find a phrase or sentence that describes, gives an example of, or states the meaning of the word. From this phrase or sentence comes a definition of the word.

Part A

Listed below are several scientific terms used in this article. Use the skill of finding word meanings from context clues to write a definition of each term. Write your definition on the blank lines provided next to each word. The page and paragraph in which the word is used is indicated in the parentheses. If the words are totally new to you and you cannot determine the meaning from the context, refer to the glossary in the back of the science textbook or to a dictionary.

1. turbine   (page 145, paragraph 4) _____

_____

2. molecules (page 146, paragraph 1) _____

_____

3. solar   (page 145, paragraph 4) (page 147, paragraph 6) _____

_____

4. silicones (page 146, paragraph 8) _____

_____

5. polymers   (page 146, paragraph 8) _____

_____

6. laser   (page 147, paragraph 5) _____

_____

Part B

Now use the terms to fill in the blanks in the following sentences.

1. The _____ system is made up mainly of the sun and the nine planets.

2. The _____ , with blades that resemble a water wheel, produces electricity.

3. _____ are used in making insulators, lubricants, and varnishes.

4. The _____ beam is used in various types of surgery.

5. _____ are composed of long chains and loops of carbon, oxygen, hydrogen, and nitrogen atoms.

6. A _____ of water is composed of two atoms of hydrogen and one atom of oxygen.

Science Reading Skill: Comparison/Contrast Pattern

In reading, writing, or speaking, comparisons are often made to point out how various things are the same as one another or different from one another. Contrasts are made to show only how things are different from one another.

The way in which material is organized often provides clues about the relationships of ideas. Material that is organized to show how things are alike or different is following a comparison/contrast pattern. When you recognize this type of pattern, your purpose in reading should be to determine what main ideas are being compared or contrasted.

Signal words and phrases such as *however, in the same way, on the other hand,* and *the difference between* are clues to help you identify the comparison/contrast pattern. Check for these clues as you practice using this skill in reading this article. Then answer the following questions using complete sentences. Do not use sentences taken directly from the article.

Part A

List five ways in which living in the 1990s was different from living in 2093.

1. _____

_____

2. _____

_____

3. _____

_____

4. _____

_____

5. _____

_____

Part B

In spite of all the differences between living in the 1990s and in 2093, certain things still remained the same. Write a brief paragraph comparing those conditions that you feel did not change during these two periods.

_____

_____

_____

_____

Answer Key

Adventures in Science

Answers for the PQRST worksheet will vary. Encourage students to use this strategy in developing good study habits. Although italics and boldface do not appear in the Science Gazettes, this is generally an important step in learning how to preview.

Issues in Science

1. T **2.** T **3.** F **4.** F **5.** F **6.** F **7.** F **8.** T **9.** T **10.** T   Written answers will vary.

Futures in Science

Part A **1.** energy-producing device; an engine whose rotating parts create electricity **2.** organic and inorganic particles that make up all substances **3.** relating to the sun **4.** new family of polymers using silicon atoms instead of carbon atons—extremely strong, lightweight, and inexpensive **5.** very long chains and loops of carbon, oxygen, hydrogen, and nitrogen atoms **6.** device that produces a very narrow and intense beam of single-color light **Part B** **1.** solar **2.** turbine **3.** Silicones **4.** laser **5.** Polymers **6.** molecule **Part A** **1.** In 2093, security, cleaning, maintenance, and communication systems were controlled by a central computer. **2.** In 2093, a combination of waste-recycling units, solar cells, and wind turbines met all the house's energy needs. **3.** People did not have elastic glass in the 1990s. **4.** Light-sensitive building materials that change color to reduce glare were also unheard of in the 1990s. **5.** Appliances and cars were much heavier in the 1990s. **Part B** Student answers will vary and could be quite inventive.

Activity Bank

TO THE TEACHER

One of the most exciting and enjoyable ways for students to learn science is for them to experience it firsthand—to be active participants in the investigative process. Throughout the *Prentice Hall Science* program, ample opportunity has been provided for hands-on, discovery learning. With the inclusion of the Activity Bank in this Activity Book, students have additional opportunities to hypothesize, experiment, observe, analyze, conclude, and apply—all in a nonthreatening setting using a variety of easily obtainable materials.

These highly visual activities have been designed to meet a number of common classroom situations. They accommodate a wide range of student abilities and interests. They reinforce and extend a variety of science skills and encourage problem solving, critical thinking, and discovery learning. The required materials make the activities easy to use in the classroom or at home. The design and simplicity of the activities make them particularly appropriate for ESL students. And finally, the format lends itself to use in cooperative-learning settings. Indeed, many of the activities identify a cooperative-learning strategy.

Students will find the activities that follow exciting, interesting, entertaining, and relevant to the science concepts being learned and to their daily lives. They will find themselves detectives, observing and exploring a range of scientific phenomena. As they sort through information in search of answers, they will be reminded to keep an open mind, ask lots of questions, and most importantly, have fun learning science.

Contents

Activity

Up in Smoke

Electrons surrounding an atomic nucleus are usually located in particular energy levels. Sometimes, however, an electron can absorb extra energy, which forces it up into a higher energy level. An electron in a higher energy level is unstable. Eventually the electron will fall back to its original position, and as it does so, it releases its extra energy in the form of light.

Different elements can absorb and release only certain amounts of energy. The amount of energy determines the color of the light that is given off. Thus the color of light given off can be used to identify particular elements. In this activity you will give extra energy in the form of heat to different elements and observe the colors given off.

Materials You Need
nichrome or platinum wire
cork
Bunsen burner
hydrochloric acid (dilute)
distilled water
7 test tubes
test-tube rack
7 chloride test solutions

Procedure ⊟ 🔥 ◉
1. Label each of the test tubes with one of the following compounds: $LiCl$, $CaCl_2$, KCl, $CuCl_2$, $SrCl_2$, $NaCl$, $BaCl_2$. Pour 5 mL of each test solution into the correctly labeled test tube.
2. Push one end of the wire into the cork and bend the other end into a tiny loop.
3. Put on your safety goggles. Hold the cork and clean the wire by dipping it into the dilute hydrochloric acid and then into distilled water. Then heat the wire in the blue flame of the Bunsen burner until the wire glows and you can no longer see colors in the flame.
4. Dip the wire into the first test solution. Place the end of the wire at the tip of the inner cone of the burner flame. Record the color given to the flame in the Data Table on the following page.
5. Clean the wire by repeating step 3. Repeat step 4 for the other 6 test solutions. Make sure you clean the wire after each test.

What You See

DATA TABLE

Compound	Color of Flame
LiCl	
$CaCl_2$	
KCl	
$CuCl_2$	
$SrCl_2$	
NaCl	
$BaCl_2$	

What You Can Conclude

1. Why do you think all of the compounds you tested were bonded to the same element—chlorine?

_____

_____

2. Why did you have to clean the wire before each test?

_____

_____

3. How do your observations compare with those of your classmates?

_____

_____

4. How can you use the flame test to identify a certain element?

_____

_____

Activity

Hot Stuff

If you look around, you will find that many of the objects and structures you depend on every day are made of metal. Metals have quite interesting and useful characteristics. One of the most important of these is the fact that metals are excellent conductors of both heat and electricity. In this activity you will find out just how well metals conduct heat.

For this activity, you will need several utensils (spoons or forks) made of different materials, such as silver, stainless steel, plastic, wood, and so forth. You will also need a beaker (or drinking glass), hot water, a pat of frozen butter, and several small objects (beads, frozen peas, popcorn kernels, or raisins).

Press a small glob of butter onto the top of each utensil. Make sure that when the utensils are stood on end, the butter is placed at the same height on each. Be careful not to melt the butter as you work with it. Press a bead, or whatever small object you choose, into the butter. Stand the utensils up in the beaker (leaning on the edge) so that they do not touch each other. Pour hot water into the beaker until it is about 6 cm below the globs of butter.

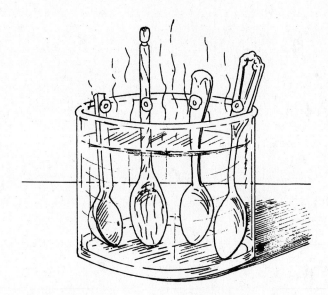

Watch the utensils for the next several minutes. What do you see happening?

Make a chart listing the material each utensil is made of, and the order in which the bead fell into the water.

Which do you expect to fall first? Which actually does?

Combine your results with those of your classmates. Make a class chart showing all of the materials used and the order in which the beads fell.

Activity

The Milky Way

Have you ever squeezed a drop of dishwashing detergent into a pot full of greasy water to watch the grease spread apart? The reason this happens has to do with the molecular structure of both the grease and the detergent. Rather than bonding together, these molecules (or at least parts of them) repel each other and move away. In this activity, you will experiment with a similar example of substances that rearrange themselves when mixed together.

What You Will Need

baking sheet or roasting pan
milk (enough to cover the bottom of the sheet or pan)
food coloring of several different colors
dishwashing detergent

What You Will Do

1. Pour the milk into the baking sheet or pan until the bottom is completely covered.

2. Sprinkle several drops of each different food coloring on the milk. Scatter the drops so that you have drops of different colors all over the milk.

3. Add a few drops of detergent to the middle of the largest blobs of color. What do you see happening?

_____

_____

■ Can you propose a hypothesis to explain your observations?

_____

_____

Activity

Chemical Reactions

Pocketful of Posies

Can you picture a meadow filled with wild flowers ranging through all the colors of the rainbow? The beautiful colors of flowers depend on combinations of chemicals carefully selected by Nature. But just as they are formed, they can also be destroyed. In this activity you will create a chemical reaction that affects flower colors.

You will need several flowers of different kinds and colors, a large jar or bottle with a lid (a clean mayonnaise jar or juice bottle will do and a plastic lid is preferable), a rubber band, scissors, and household ammonia (about 50 mL).

Procedure

1. Gather the flowers so that all of the stems are in a bunch. Use the rubber band to hold the stems together. You may have to twist it around the stems more than once.

2. Cut a large hole in the jar lid with the scissors. The gathered stems of the flowers must be able to fit snugly through the hole. **CAUTION:** *Be careful when using sharp instruments.*

3. Push the stems through the hole so that when the lid is placed on the jar, the flowers will be suspended inside the jar.

4. Pour a little ammonia into the jar—enough to cover the bottom. **CAUTION:** *Do not breathe in the ammonia vapors.*

5. Carefully place the lid with the flowers on the jar. Look at the flowers after 20 to 30 minutes. Do you observe any changes in them? If so, what do you see happening?

_____

_____

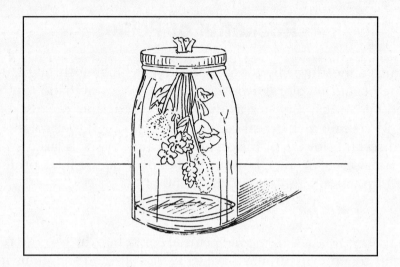

6. Compare your results with those of your classmates who may have used different flowers. Record the overall results.

Thinking It Through
▪ The pigments that give flowers their beautiful colors are present along with chlorophyll, which is green. Chlorophyll is the substance that makes photosynthesis (the food-making process in plants) possible. What do you think happened during this activity to explain your observations?

_____

_____

_____

▪ In tree leaves, colorful pigments are also present along with chlorophyll, but in this case the green chlorophyll hides the colorful pigments. What must happen to give leaves their stunning autumn colors?

_____

_____

_____

Activity

Popcorn Hop

What do you see when you pour soda or other carbonated drinks into a glass? From experience, you probably know that you see bubbles continually rising to the top—thanks to the carbonation. In this activity you will create a chemical reaction similar to the one occurring in soda and you will use it to make popcorn kernels hop!

Materials

15 mL (1 Tbsp) baking soda
45 mL (3 Tbsp) vinegar
large, clear drinking glass, beaker,
  or jar

food coloring (a few drops)
stirrer (or long cooking utensil)
popcorn kernels or raisins or mothballs
  (handful)

Procedure

1. Fill the glass container with water.

2. Add about 15 mL of baking soda, a few drops of food coloring, and stir well.

3. Drop in the popcorn kernels (or raisins/mothballs) and stir in 45 mL of vinegar. Watch the kernels for the next several minutes.

- What do you observe happening? (If the action slows down, add more baking soda.)

_____

_____

_____

_____

- Explain what you see in terms of chemical reactions.

_____

_____

_____

_____

- In an effort to preserve the natural environment, people are beginning to use Earth-friendly cleaning products. For example, rather than dumping poisonous chemicals into a sink, a mixture of hot water, vinegar, and baking soda can be used to clean drains. Why do you think this works?

_____

_____

_____

_____

Activity

Chemical Reactions

CHAPTER

2

Toasting to Good Health

Have you ever been given toast when you weren't feeling well? For some reason, toasted bread seems easier on your digestive system than untoasted bread does. In actuality, the reason is no mystery. It has to do with a chemical reaction involving the heat from your toaster. In this activity you will discover the difference between plain bread and toasted bread.

Materials
slice of white bread
slice of white toasted bread
household iodine (5 mL or 1 tsp)
drinking glass or 250-mL beaker
baking dish (or bowl with a flat bottom)
spoon (measuring spoon would be helpful)

Procedure
1. Fill the drinking glass or beaker half-full with water.
2. Mix 5 mL (about 1 tsp) of iodine into the water. Carefully pour the water-iodine solution into the baking dish.
3. Tear off a strip (about 2 cm wide) from the plain slice of bread. Dip the strip in the solution.

■ Do you observe any changes in the bread?

_____

_____

4. Tear off a strip of the same size from the toasted bread. Dip this strip in the solution.

■ Do you observe any changes in the toast?

_____

_____

■ When starch and iodine are combined, they react to form starch iodide, which is a bluish-purple. For this reason, iodine is used to test for starch. Knowing this, what can you learn from your observations?

_____

_____

■ What type of chemical reaction is involved in toasting—endothermic or exothermic?

_____

_____

■ The process of food digestion begins in your mouth. Part of this process involves breaking starches down into simpler substances. As a result of doing this activity, can you now explain why toast is sometimes recommended when you are not feeling well?

_____

_____

_____

_____

Cartoon Chemistry

Remembering the four different types of chemical reactions can be quite confusing at first. This is especially true if you rely only on memorizing complex equations of symbols and formulas. If you enjoy drawing, however, you can use your creativity to help you understand the different reactions more easily. All you need is a sheet of paper or posterboard and something to draw with—you may choose to use colored pencils, crayons, pens, or markers.

What You Do

1. Analyze the sample cartoons below.

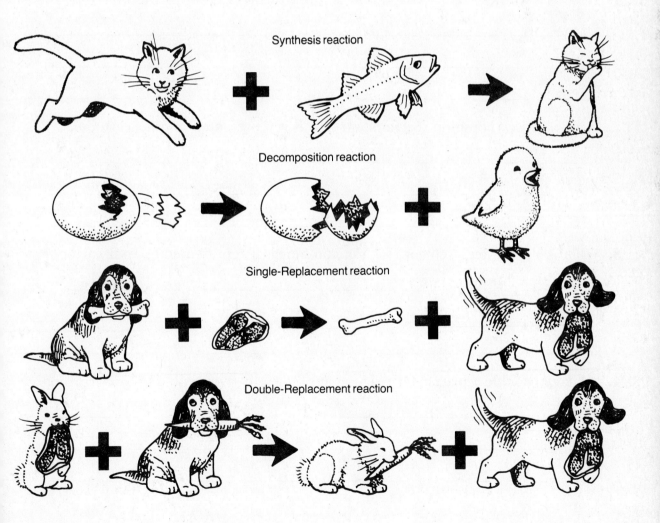

2. Describe what is happening in each frame.

a. _____

_____

b. _____

_____

c. _____

_____

d. _____

_____

3. On a separate sheet of paper or posterboard, design and draw your own cartoons to represent the four reactions. Use the sample as a model.

What You Learn

1. What is a synthesis reaction and how is it represented by your cartoon?

_____

_____

_____

2. What is a decomposition reaction and how is it represented by your cartoon?

_____

_____

_____

3. What is a single-replacement reaction and how is it represented by your cartoon?

_____

_____

_____

4. What is a double-replacement reaction and how is it represented by your cartoon?

_____

_____

_____

Activity   Families of Chemical Compounds   CHAPTER **3**

In a Jam

You have probably seen or used litmus paper to determine whether a substance is an acid or a base. But did you know that litmus paper is not the only material that can be used as an acid-base indicator? It may surprise you to learn that many foods can also do the job in a pinch! In this activity you will experiment with just such an indicator.

Materials
blackberry jam (a spoonful is enough)
warm water
small drinking glass
household ammonia (several drops)
lemon juice or vinegar (several drops)
spoon
medicine dropper

Procedure
1. Fill the drinking glass half-full with warm tap water.

2. Put a spoonful of jam into the water and gently stir it with the spoon until it is dissolved. The water-jam solution should turn a reddish color.

3. Use the medicine dropper to put a few drops of ammonia into the solution. Stir the solution once or twice. What happens to the color of the solution?

_____

_____

4. Clean the medicine dropper. Use the clean dropper to add several drops of lemon juice or vinegar to the solution. Clean the spoon and again stir the solution. What happens to the color of the solution this time?

_____

_____

Chemistry of Matter   O ■ 155

5. Compare your observations with those of your classmates who added the substance that you did not—lemon juice or vinegar. What happened to their solutions?

_____

_____

■ The jam solution is red when an acid is added to it and greenish-purple when a base is added. From your experiment, determine whether ammonia, vinegar, and lemon juice are acids or bases.

_____

_____

_____

The Next Step

Repeat the experiment several more times, each time using a different test substance. You may choose such substances as milk, juice, soda, or fruit. Be sure to clean the spoon between each stirring. Make a chart showing which substances are acids and which are bases. Combine your observations with those of your classmates.

Activity

Soda Fountain

Have you ever noticed that a slice of bread or cake is filled with tiny spaces or holes rather than being solid? These holes are made from bubbles of carbon dioxide that are produced when chemical substances, including acids and bases, in the baking ingredients react with each other. If you have ever watched someone cook pancakes on the griddle, you have seen bubbles of carbon dioxide rise to the surface as the batter heated up. You would imagine, then, that carbon dioxide bubbles have very important roles in cooking. But you may be surprised to learn that bubbles of carbon dioxide can be used to do work. In this activity, you will find out how.

Materials
plastic or glass soda bottle (1 liter or smaller)
one-hole stopper
plastic tubing to fit through the hole in the stopper
vinegar
sodium bicarbonate (baking soda)

water
large mixing container
stirrer
tissue paper, 10 cm × 10 cm

Procedure
1. In a large mixing container, make a solution that is half vinegar and half water. You need enough to fill the bottle three-fourths full.

2. Stir the solution and pour it into the bottle.

3. Place the plastic tubing through the hole in the stopper. Arrange it so that when the stopper is placed in the bottle, the tubing will extend well into the solution near the bottom of the bottle.

4. Lay the tissue paper on a flat surface. Place several milliliters of sodium bicarbonate in the center of the tissue paper. Roll the tissue paper around the sodium bicarbonate and twist it at the ends.

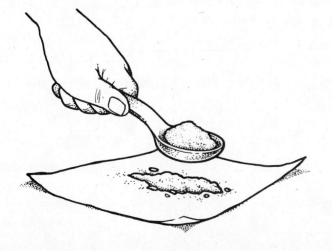

Before you go any farther, you must find a waterproof location. If possible, it would be best to complete this activity outdoors. If not, make sure you have permission to complete it at your selected location. The area directly around the activity will get wet.

5. Drop the sodium bicarbonate wrapped in tissue paper into the bottle. It will sink into the solution.

6. Quickly insert the stopper into the mouth of the bottle. Make sure the bottom of the tubing is in the liquid.

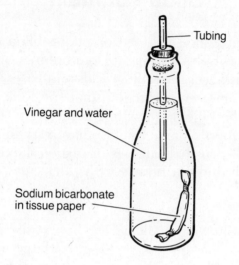

7. Stand back and watch what happens!

Observations and Conclusions

1. Describe what you observed.

_____

_____

_____

_____

2. Using the information in the chapter, determine whether vinegar is an acid or a base.

_____

3. Determine whether sodium bicarbonate, baking soda, is an acid, base, or salt.

_____

4. Using your answers to the previous two questions, explain what happens when the solution wets the tissue paper.

_____

_____

_____

_____

5. Can you hypothesize as to why the solution rises?

_____

_____

_____

_____

6. Do you know what the purpose of the tissue paper was? Explain.

_____

_____

_____

Activity

Oil Spill

You have probably seen television news reports or read newspaper articles about the devastation caused by an oil spill from a supertanker or other holding vessel. But initial reports often underestimate the full spectrum of the damage. In this activity you will simulate interactions with oil so that you can more clearly understand the dangerous consequences of an oil spill.

Materials
medicine dropper
small graduated cylinder
motor oil, used
fan
tongs
3 hard-boiled eggs, not peeled
paper towels
shallow baking pan, about 40 cm × 20 cm
white paper, 1 sheet
graph paper, 1-cm grid
beaker or jar (must be able to hold 3 eggs)

Procedure
1. Partially fill a shallow baking pan two-thirds full with water.

2. Pour the motor oil into the graduated cylinder.

3. Use the medicine dropper to remove 1 mL of oil from the graduated cylinder. Gently squeeze the oil out of the dropper into the center of the pan of water. Describe the interaction between the water and oil.

4. Mark off a region on the graph paper that is the same size as the baking pan. After several minutes, sketch the arrangement of oil in the pan of water. When you are finished drawing, count up the number of squares on the graph paper covered by oil. Remember, the area now covered was produced by only 1 mL of oil! Assuming that oil always spreads proportionately, make a chart showing the area that would be covered by 2 mL, 10 mL, 100 mL, and 1 L.

5. Place a fan beside the pan of water and oil. Turn it on and determine if the flow of air affects the spread of oil. What do you discover?

_____

_____

_____

6. Now try shaking the pan slightly. Be careful not to spill the contents. Does this reaction affect the oil at all?

_____

_____

7. Gently place the three hard boiled eggs in the jar or beaker. Pour oil into the container until it is full. Place the container under a strong light.

8. After 5 minutes use the tongs to carefully remove one egg. Remove the excess oil with a paper towel. Peel the egg. What do you observe?

_____

_____

_____

9. Remove the second egg after 15 minutes. Peel this egg and record what you observe. Remove the third egg after 30 minutes. Again peel the egg and record your observations.

The Big Picture

1. Supertankers carry millions of liters of oil. In light of your calculations, what can you say about the implications of a large oil spill?

   _____

   _____

2. What did you learn by blowing air on the oil and by shaking the water? What conditions did these procedures represent? How do these conditions affect the severity of oil spills?

   _____

   _____

3. What effect could oil have on the eggs of birds nesting near ocean water that becomes contaminated with oil?

   _____

   _____

The Domino Effect

Have you ever played with dominoes? If so, you know that dominoes can be arranged into all sorts of complicated patterns that enable you to knock them all down from a gentle tap on just one domino. Beyond playing, the falling action of dominoes can be used to represent a very complex phenomenon—a nuclear chain reaction. In this activity, you will need 15 dominoes and a stopwatch to learn more about nuclear chemistry.

Procedure

1. Place the dominoes in a row so that each one is standing on its narrow end. Each domino should be about 1–1.5 cm from the next one.

2. Gently tip the first domino in the line over so that it falls on the one behind it. You have just initiated a chain reaction.
   - What keeps the reaction going?

   _____

   _____

   - How can the row of dominoes be likened to a nuclear chain reaction?

   _____

   _____

3. Now arrange the dominoes as shown in the accompanying figure.

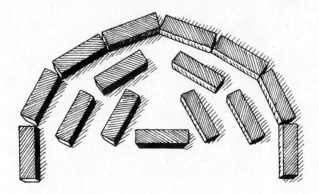

4. Gently tip over the center domino. How does this arrangement differ from the first one?

_____

_____

■ If the dominoes again represent atomic nuclei, how is this chain reaction different from the first one?

_____

_____

5. Which arrangement of dominoes do you think falls faster? Find out. Arrange the dominoes back into a single row. Use the stopwatch to measure the length of time from when you tip the first domino until the last domino falls. Record the measurement. Return the dominoes to the second pattern. Again record the time from the tip of the first domino to the fall of the last one. Which arrangement falls faster?

_____

_____

6. Now set up the dominoes as shown in the accompanying figure.

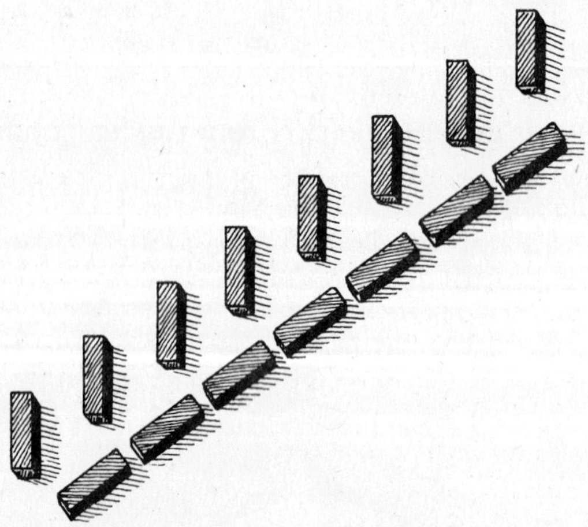

7. Tip the leftmost domino. Record the amount of time it takes for the dominoes to fall.
■ How does the length of time it takes for the dominoes to fall in this pattern compare with the times for the other two patterns?

_____

_____

■ How can you explain this arrangement in terms of a nuclear chain reaction?

_____

_____

8. Design your own arrangement for the dominoes. Determine how long it takes for all the dominoes to fall from this arrangement. Compare this time with those of the other arrangements and with those of your classmates. Make a chart or a poster showing the different arrangements and the length of time recorded for each one.

What This All Means
- Why might it be important to slow down a reaction?

_____

_____

_____

- After completing this activity, can you think of how nuclear chain reactions can be controlled in nuclear power plants?

_____

_____

_____

Answer Key

Activity: Up In Smoke

What You See LiCl: crimson; $CaCl_2$; yellow-red; KCl: violet; $CuCl_2$: blue-green; $SrCl_2$: red; NaCl: yellow; $BaCl_2$: green-yellow. **What You Can Conclude 1.** All the compounds must be bonded to the same element so that one can be sure that the differences in the observed colors are due to differences in the metals. Otherwise, one would not be able to identify the metal by the characteristic color it produced. **2.** Some of the solution from the previous test might remain on the wire. The color seen would then be a combination of the two metals. It would not be possible to isolate one from the other. **3.** All students should observe the same colors. **4.** As observed from this activity, certain elements give off characteristic colors. Given an unknown solution, a flame test could be performed to find out what color is given off. A comparison of that color with the characteristic colors recorded in the data table would identify the metal in the unknown solution.

Activity: Hot Stuff

The butter begins to melt and slide down the utensil. The butter on the metal utensil will melt first, causing the bead to slide into the water. This is because the metal is the best conductor of the heat from the water.

Activity: The Milky Way

What You Will Do 3. The colors spread out across the surface of the milk. Because the detergent spreads out, taking the color with it, it would seem that the detergent is rearranging itself so that part of its structure has the least contact possible with the milk. The reason is because part of each detergent molecule is hydrophobic and moves away from the water in the milk. The detergent spreads itself over the surface so that the hydrophobic part of each molecule can stick up into the air.

Activity: Pocketful of Posies

Procedure 5. Most of the colors disappear. Red, pink, and purple flowers turn green. White and yellow flowers stay the same. **6.** Red, pink, and purple flowers will turn green. White and yellow flowers do not change. **Thinking It Through** Students should realize that the ammonia was involved in a chemical reaction that destroyed the colorful pigments in some of the flowers and left the chlorophyll. Actually, it is the ammonia gas that rises from the liquid that destroys some of the colors but not others. When colors that hide the green chlorophyll are destroyed, the green color shows. The chlorophyll in the leaves must either be destroyed or its production must be stopped. When the chlorophyll color no longer overpowers the colored pigments, the colors of the leaves can be seen.

Activity: Popcorn Hop

Procedure 3. ▪ The popcorn kernels rise to the top of the water and then drop back down in a repeating process. ▪ The reason is that the baking soda and vinegar react to form bubbles of carbon dioxide. The bubbles attach themselves to the kernels and carry them to the surface. When the bubbles burst at the surface, the kernels sink. ▪ The baking soda and the vinegar react to form bubbles, as seen in the activity. The movement of the bubbles can lift dirt and grime off the sides of the drain pipe. When flushed with water, the dirt will be cleaned away without destroying the environment.

Activity: Toasting to Good Health

3. The strip of bread turns bluish-purple. **4.** The center of the bread will turn bluish-purple while the toasted outside remains unchanged. The untoasted part of the bread turns bluish-purple, showing that it is mostly starch. The toasted part remains unchanged. This must mean that the toasting process causes a chemical reaction that changes starch into another substance. Students who do a little research will find that the starch in the toasted areas has been changed by heat into dextrin. Dextrin iodide is not bluish-purple. Endothermic, because the bread absorbs heat from the toaster to cause starch to change into another substance. Toasting is a step in the digestion of bread. Because toast is partially digested when compared with plain bread, it is actually easier on the digestive process since the first steps have already begun and the body has to do less work in breaking it down.

Activity: Cartoon Chemistry

2. a. The cat and the fish, two separate substances, join together when the cat eats the fish. The result is one substance, a fat cat. b. One substance, the egg, breaks apart. The result is two separate substances, the chick that emerged and the cracked egg shell. c. The bone is being held by the puppy and the steak is separate. But the puppy drops the bone and picks up the steak. d. The rabbit has a steak and the puppy has a carrot. Then the puppy and the rabbit switch so that the puppy has the steak and the rabbit has the carrot. **What You Learn** *Note that the cartoons are only general models to help students understand reactions more fully.* Make sure that students realize that they in no way represent the actual events of bonding. Further, students should describe their own cartoons rather than the model. 1. A synthesis reaction is one in which separate substances, such as the cat and the fish, join together to form one substance with different properties than either substance alone. 2. In a decomposition reaction, a single substance, such as the egg with the chick, breaks down into simpler substances that have different properties than the initial substance. 3. In a single replacement reaction, two substances that are bonded together break apart and one of those substances combines with a third substance. In the case of the cartoon, the puppy breaks apart from the bone in order to recombine with the steak. Unlike the cartoon, however, the process is not reversible. 4. A double-replacement reaction is one in which two separate molecules break apart so that the components of the molecules can switch partners. In the case of the cartoon, the puppy and the rabbit switch the food they are holding in their mouths.

Activity: In a Jam

Procedure 3. The color changes to a greenish purple. 4. The color changes back to red. 5. Either substance will turn the solution red. Ammonia is a base; lemon juice and vinegar are acids. **The Next Step** Check student charts for accuracy.

Activity: Soda Fountain

Observations and Conclusions 1. When the water dissolved the tissue paper, bubbles began to form in the solution. Eventually the solution began shooting straight up through the plastic tubing in a spectacular fountain. 2. Acid. 3. Salt. 4. When the solution wets the tissue paper, the vinegar, which is an acid, reacts with the sodium bicarbonate, which is a salt. One of the results is carbonic acid, a weak acid that immediately decomposes into water and carbon dioxide gas, which can be seen in the form of bubbles. 5. The carbon dioxide bubbles rise to the surface, break, and begin to build up the pressure in the bottle. When the pressure becomes great enough, it forces the solution up the tubing and out of the bottle. 6. The tissue paper was used to delay the reaction. If the sodium bicarbonate had been poured directly into the solution, the reaction would have begun immediately. The tissue paper delays the reaction until the vinegar solution soaks through the tissue paper.

Activity: Oil Spill

Procedure 3. The oil and water do not mix. The oil spreads out across the surface of the water. 5. The moving air makes the oil disperse even further. 6. Shaking the water and oil combination should make the oil spread out further over the water. If the water is already covered, students may wish to begin again. 8. Students should be able to see that oil is beginning to seep through the shell. 9. More and more oil seeps through the longer the eggs remain in the oil. By the last egg, oil has seeped through the shell. **The Big Picture 1.** A small volume of oil, such as 1 mL or 2 mL, may fit in a medicine dropper

but when dumped into water, it quickly covers an area many times its original size. This situation is only magnified when the original volume of oil is increased. Thus an oil spill in the ocean cannot be an isolated event. The oil will spread and affect a tremendous area.
2. Blowing air and shaking the water intensified the spread of the oil. These conditions represent the winds that blow ocean waters and the waves that rock the waters. Because wind and waves are natural conditions of oceans, any oil spill will have more far-reaching ramifications than it would under ideal conditions and can be made worse when these conditions are particularly severe.
3. The baby birds forming within the eggs will be altered when the oil seeps into their food supply, which is sealed inside the egg.

Activity: The Domino Effect

Procedure 2. Each domino hits the next one until the last domino falls. The dominoes can be used to represent atomic nuclei. When the first domino is tipped, it is like an atomic nucleus being struck by a neutron. The neutron causes the nucleus to split into two smaller nuclei. When this happens at least one neutron is released. The neutron goes on to hit another nucleus. This is what is being represented each time a domino falls and hits the next one. The process continues until the last nucleus is split—or in this case, the last domino falls. **4.** In this case, each domino hits two more dominoes when it falls. In turn, the two falling dominoes tip four more over. Each falling domino results in more and more dominoes falling over in each successive row. This reaction represents the situation in which a fissioning atom releases two neutrons each time. Those two neutrons go on to strike two more nuclei. **5.** The second pattern. The first arrangement, the single line, takes longer to fall because only one domino falls at a time. **7.** The length of time is shorter than that for the straight row to fall but longer than that for the semicircular configuration to fall. Students should realize that in this configuration, eight dominoes fall without hitting other dominoes. Thus the main row of dominoes—those set up in a slanted row—represent nuclei that each releases two neutrons. One neutron splits another nuclei that releases two more neutrons. Those two neutrons hit two more nuclei. The other neutron hits a nuclei that releases another neutron. This neutron, however, does not strike another nucleus. This slows down the reaction. **What This All Means** Nuclear reactions that occur in nuclear reactors must be carefully controlled. A fast, uncontrolled reaction is not desirable. Nuclear fuel rods (dominoes) can be arranged in such a way that released neutrons are either utilized or wasted. In addition, substances that absorb neutrons can be used to prevent released neutrons from striking additional nuclei. A little research will show that some nuclear power plants submerge nuclear fuel rods in water containing boric acid. Boric acid absorbs neutrons. The rate of nuclear reactions can be controlled by altering the amount of boric acid in the water.